Towards a Sustainable

Information Society

Deconstructing WSIS

Edited by Jan Servaes & Nico Carpentier

intellect
Bristol, UK
Portland, OR, USA

First Published in the UK in 2006 by
Intellect Books, PO Box 862, Bristol BS99 1DE, UK
First Published in the USA in 2006 by
Intellect Books, ISBS, 920 NE 58th Ave. Suite 300, Portland, Oregon 97213-3786, USA
Copyright ©2006 Intellect Ltd

A catalogue record for this book is available from the British Library

Cover Design: Gabriel Solomons
Copy Editor: Holly Spradling

ISBN 1-84150-133-6

European Consortium for Communications Research Volume 2
ISSN: 1742-9420

Printed and bound in Great Britain by 4edge Ltd.

European Consortium for Communications Research

This series consists of books arising from the intellectual work of ECCR members. Books address themes relevant to ECCR interests; make a major contribution to the theory, research, practice and/or policy literature; are European in scope; and represent a diversity of perspectives. Book proposals are refereed.

Series Editors

Denis McQuail
Robert Picard
Jan Servaes

The aims of the ECCR:

• To provide a forum where researchers and others involved in communication and information research can meet and exchange information and documentation about their work. Its disciplinary focus will be on media, (tele)communications and informations research;

• To encourage the development of research and systematic study, especially on subjects and areas where such work is not well developed;

• To stimulate academic and intellectual interest in media and communications research, and to promote communication and cooperation between members of the Consortium;

• To co-ordinate information on communications research in Europe, with a view to establishing a database of ongoing research;

• To encourage, support, and where possible publish, the work of junior scholars in Europe,

• To take into account the different languages and cultures in Europe;

• To develop links with relevant national and international communication organisations and with professional communication researchers working for commercial and regulatory institutions, both public and private;

• To promote the interests of communication research within and between the member states of the Council of Europe and the European Union; and

• To collect and disseminate information concerning the professional position of communication researchers in the European region.

Table of Contents

Foreword: Towards a New Democratic Lingua Franca: Opening Speech at the ECCR WSIS conference, European Parliament March 1, 2004

BART STAES (MEP GROEN!)

The notion of the information society carries the immense hope for a better world society. In one of the more optimistic accounts – by Howard Rheingold (1993) – the newly developed information and communication technologies are said:

• to support citizen activity in politics and power,

• to increase interaction with a diversity of others

• and to create new vocabularies and new forms of communication.

From this perspective, the emancipatory and liberating aspects of ICTs will have a guaranteed impact on our languages, geographies, identities, ecologies, intimacies, communities, democracies, and economies. If we believe these utopian believers, we have finally reached the end of history, as Francis Fukuyama (in a very different analysis) wrote in 1992.

But all is not well in the new information society, and we definitely (and fortunately) have not reached the end of history.

We need to remain aware that the belief in the newness of technology and in its magical capacity to change the world has more than once led to unwarranted optimism. A nice way to symbolise this point is the following poem that sings praise over the first electronic highway: the telegraph. It was written in 1875 by Martin F. Typper, and forms a good illustration of the technological optimism that accompanied the introduction of the telegraph.

> Yes, this electric chain from East to West
> More than mere metal, more than mammon can
> Binds us together – kinsmen, in the best,
> As most affectionate and frankest bond;
> Brethren as one; and looking far beyond
> The world in an Electric Union blest!

When dealing with the present-day information society we should – as always – remain sceptical towards all forms of technological determinism and economic reductionism. ICTs have created a number of opportunities that we urgently need to exploit to their full capacity. They also have created a number of new problems, dysfunctions and distortions, which evenly need to be addressed urgently.

In short, technologies are only as good as the people that put them to use.

One of these problem areas that have captured our imagination has been called the digital divide. While the reduction of the differences in access to ICTs – both in Europe and at a global level – remains of crucial importance, we should keep in mind to include an emphasis on user skills, user needs and on content that is considered relevant by the users. Furthermore, we should also keep the societal context in mind: digital exclusion should remain strongly connected to the much broader phenomena of social and economic exclusion and poverty.

And social and economic exclusion (which includes digital exclusion) cannot be reversed without tackling the plurality of factors that leads to inequality. Creating access to ICTs is indeed one of the many tools for societal improvement but should be embedded in a more general perspective on inclusion, development and poverty reduction.

Moreover, access is not the only problem that puts a shadow over the information society's realisations. Here I would like to refer to Oscar Gandy's article in the *Handbook of New Media* (2002). In this article, which has the following title *'the real digital divide: citizens versus consumers'*, he sees *'the new media as widening the distinction between the citizen and the consumer.'* (Gandy, 2002: 448) His main concern is that the 'new economy' will incorporate and thus foreclose the democratic possibilities of the new media. He continues by predicting that the balance between both models will eventually determine the role of ICTs (and more specifically of the Internet) in post-industrial democracy.

This prediction creates a serious challenge and requires a partial reorientation of our attention. The (democratic) needs of citizens as part of a wide range of diversified users communities should be taken more into account. This implies a more user – and needs-oriented approach that does not detach technological and economic development from the democratic society in which it takes place.

We lose too many opportunities to strengthen and deepen our democracies when we reduce ICT users to their role of consumers of commercial and government services. We also lose too much when we forget that we are living in an information and communication society, and not just in an information society. In other words, we should avoid that 'information' becomes our new fetish, but instead try to discover how policies can support and stimulate a sustainable and democratic dialogue in Europe and in the world.

In short, more than ever before, we need to put citizens, and not technology, first. When the United Nations' General Assembly adopted a resolution that (among other things) asked for the active participation by non-governmental organisations in the World Summit on the Information Society (WSIS), the stakes were high. The usually inaccessible arena of inter-state negotiations, at least partially, became accessible for civil society and business actors. Before, civil society was usually seen marching in protest, outside the summit location, a situation that is symbolised by the name of that one American city: Seattle.

In contrast to this exclusionary approach, the World Summit on the Information Society was announced as a major step forward regarding citizen participation. In one of the EU documents for the Preparatory Committee Meeting the summit itself is even seen as a model for the future role of civil society.

After the Geneva summit the disappointment of civil society actors can hardly be underestimated. I'd like to quote from their *Civil Society Declaration to the World Summit on the Information Society*, which is called *Shaping Information Societies for Human Needs* (2003). The civil society representatives have agreed unanimously upon the following statement: '*At this step of the process, the first phase of the Summit, Geneva, December 2003, our voices and the general interest we collectively expressed are not adequately reflected in the Summit documents.*'

When I questioned Commissioner Erkki Liikanen on this matter, and on his plans towards stimulating and increasing citizens' participation in the next phases of the WSIS, Liikanen expressed his appreciation for the involvement of civil society organizations in the process leading to the summit and in the summit itself. Despite the fact that (according to Commissioner Liikanen) the WSIS remains an intergovernmental summit, within the framework of the United Nations, he has witnessed the growing emergence of a lingua franca between governments and their civil societies.

Our information society is indeed in need of a lingua franca that respects the cultural diversity in and outside Europe; that creates a new balance between Europe and it citizens, and that strongly situates Europe in a more free, peaceful and just world.

References

European Union (EU). 2002. *The UN World Summit on Information Society. The preparatory process.* Reflections of the European Union (WSIS PrepCom1 document 19/6/02), Brussels: EU, accessed 15/11/2004, http://europa.eu.int/information_society/topics/telecoms/international/wsis/eu_paper_fin%20_en_19jun02.pdf.

Fukuyama, Francis. 1992. *The End of History and the Last Man*. New York: Free Press.

Gandy, Oscar. 2002. 'The real digital divide: citizens versus consumers', pp. 449–460 in L. Lievrouw & S. Livingstone (eds.) *The Handbook of New Media*. London: Sage.

Rheingold, Howard. 1993. *The Virtual Community. Homesteading on the Electronic Frontier.* Reading, Massachusetts: Addison-Wesley.

WSIS Civil Society Plenary. 2003. *Civil Society Declaration to the World Summit on the Information Society.* Shaping Information Societies for Human Needs, http://www.itu.int/wsis/docs/geneva/civil-society-declaration.pdf.

Introduction: Steps to Achieve a Sustainable Information Society

JAN SERVAES & NICO CARPENTIER

The Information Society is an evolving concept that has reached different levels across the world, reflecting the different stages of development. Technological and other change is rapidly transforming the environment in which the Information Society is developed. The Plan of Action is thus an evolving platform to promote the Information Society at the national, regional and international levels. The unique two-phase structure of the WSIS provides an opportunity to take this evolution into account. (Plan of Action, WSIS Conference, Geneva, December 2003)

From Information Society to Knowledge Societies

Though many authors express serious doubts about the validity of the notion of an Information Society (IS), a variety of criteria could be used to distinguish analytically definitions of the IS. Frank Webster (1995: 6), for instance, identifies the following five types of definitions: technological, economic, occupational, spatial, and cultural. The most common definition of an IS is probably technological. It sees the IS as the leading growth sector in advanced industrial economies. Its three strands – computing, telecommunications and broadcasting – have evolved historically as three separate sectors, and by means of digitization these sectors are now converging.

Throughout the past decade, however, a gradual shift can be observed in favor of more socio-economic and cultural definitions of the IS. The following definition, drafted by a High-Level Group of EU experts, incorporates this change:

The information society is the society currently being put into place, where low-cost information and data storage and transmission technologies are in general use. This generalization of information and data use is being accompanied by organizational, commercial, social and legal innovations that will profoundly change life both in the world of work and in society generally. (Soete, 1997: 11)

Others prefer to use the term Knowledge Societies (in plural) for at least two reasons: (a) to indicate that, depending on historical and contextual circumstances, there are more roads than just one to a future Knowledge Society, and (b) to clarify the shift in emphasis from information and communication technologies (ICTs) as 'drivers' of change to a perspective where these technologies are regarded as tools which may provide a new potential for combining the information embedded in ICT systems with the creative potential and knowledge embodied in people. '*These technologies do not create the transformations in society by themselves; they are designed and implemented by people in their social, economic, and technological contexts.'* (Mansell & When, 1998: 12) At the same time Williams' (1999: 133) wise remark should be kept in mind: *'While we have to reject technological determinism, in all its forms, we must be careful not to substitute for it the notion of a determined technology.'*

Global Regulatory Frameworks

Global change and developments in ICTs are affecting practices of political conduct at all levels of society. In an increasingly globalized and regionalized European Union, politicians formulate their IS policies within an international and global framework, with national interests at stake. The first phase of the World Summit on the Information Society (WSIS), that took place in Geneva in December 2003, added another dimension to the already complex dynamics of global IS governance. The decade preceding WSIS was marked by a number of radical initiatives toward bringing ICT regimes increasingly outside the national domain, as the 1990s was marked by the European Union's and the United States' vigorous telecommunications and IS policies. Broadcasting, telecommunication and information policies are now converging at a European and worldwide level, along side technological and economic convergence. In this regard it is worth referring to the 1993 agreement, signed by 130 countries with the World Trade Organization (WTO), in which communication was treated as a service. This was a major milestone on the road toward an internationalized communications system. The 1997 *Agreement on Basic Telecommunications Service*, signed by 69 countries, set the tone for the opening up of domestic markets to foreign competition. At the European level, this has been made explicit in the *Green Paper on the Convergence of the Telecommunications, Media and Information Technology*, published in 1997, and its follow-up, the 2003 *Regulatory Framework for Electronic Communications Networks and Services*. The latter clearly indicates the EU approach, which is that all communications should be regarded as part of the same regulatory concept. Viviane Reding, the new European Commissioner for the Information Society and Media, confirmed this as follows: *'European Audiovisual Policy has consistently sought to provide a framework favorable to the development of the audiovisual sector and to support the transnational dimension of this essentially cultural industry. In this respect the* Television without frontiers *Directive is the essential centerpiece for a 'business without frontiers' drive. This is as true today as it will be in the future in a wider media perspective. The 'leitmotiv' is to create added-value at European level and not to seek to do what can better be done at national level.'* (Reding, 2004: 2) In her contribution to this book Barbara Thomas develops this argument further by arguing in favor of an inclusion of the Public Service Broadcast (PSB) philosophy in the WSIS agenda. Her analysis also points to the potential cross-fertilization of the PSB's current efforts with the WSIS agenda when discussing vital elements like increased access, capacity building and education, and support for cultural identities and diversity.

Our Objectives

The above issues were explored and discussed in the first volume of the European Consortium for Communication Research (ECCR) Book Series (Servaes, 2003). This second volume is equally ambitious. It presents some of the papers presented during the March 2004 ECCR conference on the evaluation of the first phase of the WSIS, plus a number of additional chapters written by ECCR members.

One specific point of attention, both during the ECCR conference as well as in this book, is the assessment of the inclusion and role of civil society and other so-called stakeholders in the decision-making process, as Resolution 56/183, adopted by the 90th plenary meeting of the General Assembly on 21 December 2001, called for their active participation in the WSIS. In one of the EU's Preparatory Committee (PrepCom) documents, the summit itself was seen as a model for the future role of civil society (and commerce): *'The preparatory process is almost as important as the political outcomes of the Summit itself. The format and positioning of the Summit will be key factors for an event which will attract attention and activate a decentralized follow-up process, not only at political level but also in society at large.'* (EU, 2002: 12)

This touches directly on the second objective of the ECCR conference and this book. Since the Lisbon summit in March 2000, official texts of the European Commission teem with new terms coined with reference to the IS, such as e-Europe, New Information and Communication Technologies (NICTs), online world, knowledge and innovation economy, etc. The very same ambitions that are present in the WSIS discourse can be found at the European level. In the *eEurope 2002 Plan of Action* (EU, 2001: 4), for instance, a call towards the member states can be found in order to *'draw the attention of citizens to the emerging possibilities of digital technologies to help to ensure a truly inclusive information society. Only through positive action now can info-exclusion be avoided at the European level.'* As Bart Staes remarks in his foreword, these ambitions are not only situated at the level of merely creating access to ICTs. Equally important are the democratic needs of these citizens that are embedded in a diversity of user communities. The potential of ICTs to stimulate a democratic dialogue amongst them, however difficult this ambition is, should not be discarded. An example of this potential can be found in Stefano Martelli's article on the *Telematic Portal for the communication of the 'third sector' in Palermo*, which aims to visualize their pro-social activities.

This book's general objective is to analyze and evaluate these different ambitions. But, at the same time, we want to present a number of recommendations for consideration to policy-makers and researchers, which could contribute to a sustainable agenda for the future IS.

Towards a Sustainable Information Society

While sustainability was initially formulated in terms of environmental preservation, the sustainability debate has broadened its scope to include social, economic and cultural aspects. What sustainability is all about has also changed – from static views that emphasize the preservation of current resources for future generations towards more dynamic views, which emphasize the development of greater opportunities for future generations. Therefore, today a more multidimensional view on sustainability is being presented which implies a holistic and integrated policy framework of environmental compatibility, economic stability, social sustainability and cultural diversity. As Peter Johnston (2000: 9) argued, '*on each of these issues, three aspects are important. Firstly, our 'understanding' of the risks and opportunities for action; secondly, the commitment of key organizations to work together to maximize benefits and minimize risks; and finally, raising public awareness, not only to ensure democratic support for appropriate policy measures, but to engage every citizen in the 'life-style' changes that may be necessary for effective change.*'

Therefore, the key conclusion in Johnston's afterword to this book is that investment in ICTs must be accompanied by investment in skills and organizational change. Here he argues for a more systemic approach to development of a sustainable IS: greater synergy between RTD, regulation and deployment actions; greater investment in more effective public services, notably for health care and education, as well as for administrations; and more active promotion of 'eco-efficient' technologies and their use.

Whether ICTs will, in the end, contribute to sustainability or not, essentially depends on the further development of global environmental, economic, social, democratic and cultural governance frameworks and corresponding attitudes and values. Matthias Fritz and Josef Radermacher (2000: 57) argue that '*the design of these frameworks determines whether they will lead to new, resource efficient lifestyles and working methods which make use of advanced IS Technologies and improve the quality of life significantly in all world regions, i.e. encouraging tele-working, electronic commerce or life-long learning. Building such frameworks is the single most important challenge to policy, industry, research and the civil society when entering the 21st century.*' Therefore, the transition to a fully sustainable IS requires one critical ingredient: collective positive action to shape it. The WSIS was an attempt to create such a moment at the global level.

Positioning WSIS

The proposal to host the WSIS was endorsed by the Council of the International Telecommunication Union (ITU) at its 2001 session. The 90th General Assembly of the United Nations (UN) officially adopted the proposal on 21 December 2001 as Resolution 56/183.

The General Assembly recognized '*the urgent need to harness the potential of knowledge and technology for promoting the goals of the United Nations Millennium Declaration (Resolution 55/2) and to find effective and innovative ways to put this potential at the service of development for all.*' In other words, the aim of the WSIS is to develop a global framework to deal with the challenges posed by the IS.

The WSIS differs from other UN conferences in that it is a two-phase process culminating in two 'world summits', the first one took place in Geneva from 10–12 December 2003, with the second to be held in Tunis from 16–18 November 2005.

More importantly, and again in contrast to previous UN conferences, the General Assembly placed a strong emphasis on the participation of non-state actors, as they encouraged '*effective contributions from and the active participation of all relevant United Nations bodies, in particular the ICT Task Force, and encourages other intergovernmental organizations, including international and regional institutions, non-governmental organizations, civil society and the private sector to contribute to, and actively participate in, the intergovernmental preparatory process of the Summit and the Summit itself.'* (Resolution 56/183) The idea was that the deliberations at the WSIS should be of a consensual nature, incorporating the viewpoints of multiple actors. This has since become known as the so-called multi-stakeholder approach.

The Multi-Stakeholder Approach

Despite the problematic issues inherent in the WSIS initiative, the novelty and significance of the program stems from the fact that WSIS was the first international event bringing together multi-stakeholders-governments, civil society, private interest groups and bureaucrats – from all over the world to reflect on the future of IS from a people-centered, human rights perspective: a perspective which is lacking in current national and supranational policies. This new multi-stakeholder approach, and especially the role and participation of civil society (including researchers and academics), is extensively analyzed and debated in this volume (in the contributions by Bart Cammaerts & Nico Carpentier, Divina Frau-Meigs, Stefano Martelli, Claudia Padovani & Arjuna Tuzzi and Ned Rossiter).

Civil Society

Civil society is traditionally defined in opposition to the state. The values of civil society – 'civility', respect for individual autonomy and privacy, trust amongst peoples, removal of fear and violence from everyday life, etc. – operated as a counterpoint to the rules and purposes of the state whose centralized political authority administered the lives of people within a given territory. Many have argued that the mutually constitutive relationship between the state and civil society has been eroded with the advent of globalized economies, 'flexible accumulation' and the abstraction of social and cultural relations that attend NICTs. Others have suggested that the notion of 'civil society' should be abandoned due to its universalization of European values.

Although civil society was an integral part of the preparatory process, the collaboration was not always smooth and easy. By responding fast and to the point, with professionalism and expertise, civil society organizations (CSOs) had to earn the respect of initially hostile or skeptical nation-states. In their article Claudia Padovani and Arjuna Tuzzi point to the importance of civil society's external and internal negotiative capacities. CSOs – for instance those in the human rights and gender caucuses – showed that they were capable of working with governments. At the same time these NGOs, grass roots groups, activists and many other organizations and individuals proved themselves capable of setting up an internal dialogue. Divina Frau-Meigs elaborates this argument in her chapter, as she points to the increased legitimacy of the role of NGOs within the ranks of other civil society actors.

As a result of this capacity for arguing and for implementing a soft-yet-firm civil disobedience, which did not balk at intense lobbying with the official representatives of supportive nation-states, some gains were obtained. Most observers agree that civil society has positioned itself as a structuring, pacifying as well as constructive power.

Nevertheless, most authors in this book do not turn a blind eye to the problems related to this 'new' form of global governance. Bart Cammaerts and Nico Carpentier scrutinize the power balances between state actors, intergovernmental organizations and civil society. Without disregarding the novelty of the shift towards more equal power balances, they conclude that 'extended consultation' might be a more accurate description than 'participation'.

From a slightly different perspective Ned Rossiter argues that notions of civil society persist within an era of *informationality*. He suggests that organized networks and their use of ICTs invite a rethinking of civil society–state relations. The WSIS is considered as a temporary supranational institution through which civil society has established a new scale of legitimacy, albeit one that must now undergo a process of re-nationalization and re-localization in order to effect material changes. Rossiter proposes that organized networks – as distinct from networked organizations – are the socio-technical form best suited to address the complex problematic of multi-scalar dimensions of informational governance. In doing so, he raises doubts about the extent in which the multi-stakeholder approach can go *'beyond some of the tenets of Third Way politics'*.

This risk of incorporation places CSOs in an awkward position, as they have to engage in complex multi-scale negotiations and dialogues, both external and internal to civil society, but still find themselves trapped in unequal power relations and in the position to defend specific values that might conflict with the processes and outcomes of *realpolitik*. One of the advantages of these 'organized networks' as Rossiter calls them or 'rhizomes' as they are termed in Michel Bauwens' article, is their mobility, contingency and elusiveness. This feature makes them capable of using a Janus-head strategy, combining strategic and partial incorporation with continued resistance and independent critique.

Networks and Communities
Because of the importance of these networked social spheres, two articles in this book provide us with an in-depth analysis of on the one hand so-called virtual communities and the other P2P (peer-to-peer) networks.

Paul Verschueren reviews the concept of the 'virtual community' from three different angles. Firstly, it considers the virtual community in early utopian and dystopian discourses. Secondly, it deals with the electronic field studies that focus on virtual communities as interactional fields. Thirdly, it shows how research interest is shifting away from the virtual community as a bounded unit of social interaction towards a much broader, contextual and everyday life perspective.

Michel Bauwens explores the potential of P2P networks. Peer-to-peer is a specific form of a network, which lacks a centralized hierarchy, and in which the various nodes can take up any role depending on its capabilities and needs. P2P is an 'egalitarian' network, a form of 'distributive and cooperative intelligence'. Thus, intelligence can operate anywhere, and it lives and dies according to its capacities for cooperation and unified action. He relates it to Alan Page Fiske's typology in that P2P particularly 'reflects' and 'empowers' two particular forms of sociality: 'Equality Matching' and 'Communal Shareholding'.

Not All State and Civil Society Actors are Alike

As Sassen (1996) puts it, global processes materialize in national contexts. It is important, then, to understand the role of distinctive national forces and patterns in the context of globalization, regionalization and localization.

The point of departure for Miyase Christensen's case study of Turkey is the contention that telecoms infrastructure and the social shaping of national policy rhetoric constitute the building blocks toward the emergence of an IS in any context. At their current stage, telecoms policy and IS regimes in Turkey, a candidate to the EU, are shaped, first and foremost, by the binding policies of the EU and by Turkey's own national power geometry. The role of the newly flourishing civil society in Turkish policymaking remains minimal.

Despite the recent liberalization of the Turkish telecoms market in January 2004, as was pressed by the EU Commission, and despite Turkish efforts, marked by such initiatives such as *eTurkey*, to catch up with the EU's supranational policy context, Christensen proposes that national specificities in the form of institutional structures and power relations are the primary determinants that shape the IS in EU candidates such as Turkey today. During the WSIS, the Turkish participation did no go beyond the official national agenda, its emphasis on economic development and its lip service to social issues. Christensen also shows how in the Turkish case the rhetorics of access to information and knowledge become intertwined with the call to fight terrorism.

Absent Others

The emphasis on civil society might give the impression that the multi-stakeholder approach was limited to state and civil society actors. In contrast to this impression, business actors were indeed explicitly included in the calls for multi-stakeholder participation. A major disappointment, however, was the low level of private sector participation. Small and medium-sized enterprises (SME) were hardly represented, and only a number of organizations linked to multinationals attended. Cammaerts & Carpentier counted only 28 CEOs who attended WSIS. The only 'big' industry players, within the information technology sector, who did send their CEO to the WSIS were Eutelsat (France), Nokia (Finland), Oracle (US), Fujitsu (Japan), Siemens (Germany) and Vodaphone (UK).

Also absent were the news media. This is even more surprising, since a number of important issues discussed at the WSIS, such as freedom of expression and freedom of the press, are often considered crucial by these media.

Declaration and Action Plan

The first phase of the WSIS in December 2003 ended with the adoption of two official documents: a *Declaration of Principles* and a *Plan of Action*. Controversial issues such as ICTs financing in the South and Internet Governance were debated during the preparatory process, but no agreements could be reached on them. They were left out in Geneva and are to be re-examined in the second phase of the summit in Tunis. Two working groups will examine issues on *Internet governance* and the creation of a *Digital Solidarity Fund* proposed by Senegal as a financial mechanism for ICTs in Southern countries.

Digital Solidarity Fund

Information and communications infrastructure is an essential foundation for an inclusive IS. Despite the existence of national universal service mechanisms, its construction is a task for which many countries, not only developing countries, require international cooperation.

The WSIS *Declaration of Principles* calls for digital solidarity and establishes the creation of a Digital Solidarity Fund, which, to be effectively operational, given the failure of many programmes based on principles of equity, requires convincing possible donors of the existence of 'other' interests.

The presence of network externalities in advanced telecommunication services and the role of telecommunications as a tool for the provision of global public goods (knowledge dissemination, economic development) are the factors proving that the advantages of this programme would not be restricted to the recipient countries. These issues are further questioned and explored in the chapter by Claudio Feijóo González, José Luis Gómez Barroso, Ana González Laguía, Sergio Ramos Villaverde & David Rojo Alonso.

Internet Governance

The working group on Internet governance has four main tasks:

• To develop a working definition of Internet governance;

• To identify the public policy issues that are relevant to Internet governance;

• To develop a common understanding of the respective roles and responsibilities of governments, existing intergovernmental and international organizations and other forums as well as the private sector and civil society from both developing and developed countries;

• To prepare a report on the results of this activity to be presented for consideration and appropriate action for the second phase of WSIS in 2005.

There are two general strands in defining Internet governance. One centered on the governance *of* the Internet, which basically accounts for the technical mechanisms and generally focused on the operations of ICANN. Governance *on* the Internet meanwhile covers a broader range of issues such as pricing, interconnection, network security, cyber crime, spam and others. The ITU, along with other organizations such as the Organization for Economic Cooperation and Development (OECD), International Consumer Protection and Enforcement Network, has been working on some of these issues.

Another area of concern relates to the very management of the Internet. The domain name system is basically controlled by the US Department of Commerce. This poses major sovereignty questions. Developing nations prefer to have an international agency, such as the ITU, govern the Internet rather than continue with the current arrangement with the US Department of Commerce.

Also other politically sensitive issues, such as intellectual property rights, trade of goods and services and debt release were hardly addressed. Delegations of

Northern countries (the United States, in particular) put a lot of effort in keeping them out of the WSIS agenda, arguing that it was not the appropriate forum to address them.

In sum, some people believe that the *Declaration of Principles* and the *Plan of Action* are too technical and have not succeeded in introducing the real social aspects, such as the human face of globalization, education in the IS, etc.

Tunis 2005

Due to the difficulties faced to reach 'strong' agreements in the first phase of the WSIS and lack of a clear leadership, the Tunis phase of the WSIS has had a hard time to start. In June 2004 the first Preparatory meeting of the second phase was held in Hammamet, Tunisia. This meeting was dominated by a heated debate on issues of human rights and freedom of expression in Tunisia. One finally agreed that the focus of the preparatory process to the Tunis phase should be two-pronged: (a) it should provide solutions on how to implement and follow up the Geneva decisions by stakeholders at national, regional and international levels with particular attention to the challenges facing the least developed countries; and (b) it should complete the unfinished business in Geneva on Internet governance and Financing. The reports of the Task Force on Financing mechanisms and the report of the Working Group on Internet governance would provide valuable inputs to the discussion. A consensus was also obtained that the agreements reached in the Geneva phase should not be re-opened.

Important Issues for the Future

Which are the important issues left for a discussion on the future sustainable IS? Before taking a more general stance, explicit emphasis has to be placed on what lies at the heart of the ECCR: research and education. In her article, Divina Frau-Meigs assesses the renewed place of research in the development of possible Knowledge Societies. She emphasizes the need to increase the social dimension of ICT policies, to develop new forms of awareness raising activities, to support cross-country research and to re-formulate the economic drivers of the digital growth. These points are also stressed in the ECCR afterword.

Below we provide the following (not exhaustive) list of topics, which are being further detailed and discussed in this book.

- Freedom of expression and the respect of human rights;

- Communication rights;

- Cultural and linguistic diversity, as for instance articulated in Unesco's *International Convention on the Protection of the Diversity of Cultural Contents and Artistic Expressions*;

- Access to the WWW, often affected by the respect for rights and multilingualism, also remains an unavoidable issue, if a true IS is to become a reality;

- Internet governance as well as intellectual property issues are at the heart of the debate;

- Lifelong education for the Knowledge Societies of the future. Education must be given more attention since it is central to the use of technology;

- More fundamental academic research is needed to perform realistic and non-commercial assessments and recommend social solutions for a sustainable future;

- More groups and individuals should be invited to participate in the second phase, such as small and medium-sized enterprises (SMEs), and Open Source and Free Software groups.

References

European Union (EU). 2002. *The UN World Summit on Information Society. The preparatory process.* Reflections of the European Union (WSIS PrepCom1 document 19/6/02), Brussels: EU, accessed 15/11/2004 from http://europa.eu.int/information_society/topics/telecoms/international/wsis/eu_paper_fin%20_en_19jun02.pdf.

Fritz, Matthias, Radermacher, Josef. 2000. 'Scenario Modelling and the European Way', IST (2000), *Towards a Sustainable Information Society*. Report of the Conference on 21–22 February 2000, EU, Brussels.

Johnston, Peter. 2000. 'Towards a sustainable Information Society', IST, *Towards a Sustainable Information Society*. Report of the Conference on 21–22 February 2000, EU, Brussels.

Mansell, R., When, U. (ed.)1998. *Knowledge Societies. Information Technology for Sustainable Development*. Oxford: Oxford University Press.

Mattelart, A. 2001. *Histoire de la société de l'information*. Paris: La Decouverte.

Negroponte, N. 1995. *Being Digital*. London: Hodder and Stoughton.

Reding, Viviane. 2004. *Business without frontiers: Europe's new broadcasting landscape*, Paper: Europe Media Leaders Summit, London, 7 December 2004.

Sassen, S. 1996. *Losing Control? Sovereignty in an Age of Globalization*. New York: Columbia University Press.

Servaes, Jan (ed.) 2003. *The European Information Society: A reality check*. ECCR Book Series, Bristol: Intellect Books.

Slevin, J. 2000. *The Internet and Society*. Cambridge: Polity Press.

Soete, L. 1997. *Building the European Information Society for us all. Final policy report of the high-level expert group.* Brussels: EU-DGV.

Webster, F. 1995. *Theories of the Information Society*. London: Routledge.

Williams, Raymond. 1999. *Television: Technology and Cultural Form*. London: Routledge Classics.

1: The Unbearable Lightness of Full Participation in a Global Context: WSIS and Civil Society Participation

BART CAMMAERTS & NICO CARPENTIER

Introduction

Global politics remained up until a few decades ago a restricted area, which was mainly accessible to nation states, and more specifically to the dominant superpowers at that time in history. This has gradually changed and 'new' actors have emerged on the global scene. One concrete manifestation of this is that states and international institutions have adopted a so-called 'multi-stakeholder approach'[1] to global and regional governance, involving more and more business – as well as civil society-actors. The rhetoric that surround these alleged inclusionary practices tend to make use of a very fluid signifier: participation. It is now claimed more and more that civil society, as well as business actors, are 'participating' in the global political processes that build future societies. This chapter asserts that these rhetorics are discursive reductions of the plurality of meanings that are embedded in the notion of participation. By confronting these rhetorics on participation with the organizational practices related to a world summit, more specifically the World Summit on the Information Society (WSIS) and its preparatory meetings called PrepComs, we will be able to show that a specific and reductionist definition of participation is produced, which excludes the possibility of a series of more balanced power relations. This analysis illustrates at the same time the problems encountered when (optimistically unprepared) introducing the notion of participation in processes of regional or global governance. It will also show that power remains an important concept that often gets obscured or masked. By making these implicit and explicit power mechanisms visible this chapter would like to contribute towards the evaluation of participatory practices within global settings.

We will be focussing foremost on civil society and its role within the WSIS process.[2] Civil society is a notion that has seen its respectability increased in academic, as well as policy discourses. After the fall of the Berlin Wall, Eastern European civil society organizations received recognition for their role in the democratization of Eastern Europe. And new (and old) social movements in the West were (again) seen as carriers of life, sub- or identity politics. Like many of these concepts the exact meaning of civil society is of course contested. Without going too much into this debate, we adopt a Gramscian perspective in this regard, making an analytical distinction between the state, the market and civil society, as a relatively independent non-profit sphere in between market and state, where (organized) citizens interact. This does, however, not mean that we see civil society as one singular actor. Civil society is diverse in its structures, going from grass roots to regional or international civil society organizations. It is also diverse in its ideological orientations, going from extremely conservative to radically progressive. While this chapter addresses the power mechanisms at play *between* civil society actors, states and international organizations in terms of access and participation to the WSIS process, the power relations and mechanisms *within* civil society, states and international organizations are unavoidably black boxed for analytical reasons, without however denying their existence. Furthermore, it has to be noted that speaking of a 'global' civil society is still contested and questions can and should be raised concerning the representativeness of civil society actors active within a global context (including the WSIS).

Before addressing the notion of participation and power and applying this to the case of the WSIS, we need to place consultation of and participation by civil society actors in an historical perspective.

Historical Contextualization of the Participatory Trajectory

One of the concrete results of the globalization processes from the 1990s onwards, was the recognition that nation states were no longer the only players on the international stage (Rosenau, 1990; Zacher, 1992; Sassen, 1999). Civil society actors as well as business actors have manifested themselves increasingly as legitimate actors in processes of global governance. At the same time the number of issues requiring global solutions also increased and became more prominent on the political agendas of citizens, civil society organizations and (some) governments (Urry, 2003; Held, et. al, 1999: 49–52; Beck, 1996). Examples of such issues are child labour, ecology, terrorism, crime, mobility, migration or human rights. In this regard we can also refer to the emergence of transnational notions of citizenship (Van Steenbergen, 1994; Bauböck, 1994; Hauben, 1995; Hutchings & Dannreuther, 1999; Sassen, 2002). This does, however, not mean that transnational issues or transnational networks as such are a totally new phenomenon as Boli and Thomas (1997: 176) have shown in their historical analysis of non-governmental organizations. In this regard there can also be referred to the Socialist International or the Suffragette movement (Geary, 1989; French, 2003). But it is fair to say that the scope and degree of cosmopolitanism has drastically increased in recent decades (Vertovec & Cohen, 2002).

Another observation relates to a crisis of institutional legitimization, be it on the level of the nation state or international/regional organizations. States are caught between the possible and the desired: they have to operate within strict budgetary and legal frameworks, international obligations and co-operative regimes and are at the same time confronted with citizens' high demands, national interests and cultural specificities. International organizations partially build on the legitimacy of their member states, but the more the representative democratic system at the national level is being questioned and debated, the more difficult it has become for international institutions to solely rely on state representatives to formulate policies. In a world of multi-level governance, international organizations also desperately need democratic legitimization in their own right (Schild, 2001), which is often of a highly questionable nature.

The recent rise in (global) political discourses of notions such as multi-stakeholder governance also has to be seen against the backdrop of theoretical efforts to extend the democratic principles to the realm of global politics. In this regard there can be referred to David Held's conceptualization of cosmopolitan democracy and to its 'realist' critical responses (Held, 1995/1997; Hutchings & Dannreuther, 1999; Saward, 2000; Patomäki, 2003). Ideas like the instalment of a democratic world parliament and government, as put forward by Held, are burdened with so many constraints that it is highly questionable whether they will materialise in the foreseeable future or indeed ever, given the complexity of the world system and the lack of – or defuncts in – democracy in many national contexts. Patomäki (2003: 371) points to many of these constraints and argues for the conceptualization of a global democracy *'in contextual and processual terms, by revising social frameworks of meanings and practices by means of cumulative but contingent and revisable reforms, also to induce learning and openness to change, in the context of cultivating trust and solidarity'*, rather than a closed linear process towards cosmopolitan democracy. From Patomäki's perspective, multi-stakeholder processes are a step in a learning process of all actors involved to build trust and to gradually reform and democratise international politics.

International institutions such as the EU and the UN look increasingly to civil society and business actors to legitimise policies that can build on the broadest support possible from the different actors involved in the complex game of governance. In this regard business actors have become crucial partners, as nation states are no longer active economic actors and are restricted budget-wise by the international financial markets or regional agreements such as the European Monetary Union. At a rhetorical level, civil society is then perceived as representing the local grass roots-level, specific interest groups, transnational social movements, counterbalancing the dominance of corporate actors and still, to a large extent, also of state representatives.

Forging links with civil society organizations has for many years also been a strategy of UN institutions in order to increase transparency and accountability when taking global initiatives. In fact, the consultation of civil society is even embedded in the 1945 Charter of the UN. Article 71 of the UN Charter states:

> *The Economic and Social Council may make suitable arrangements for consultation with non-governmental organizations which are concerned with matters within its competence. Such arrangements may be made with international organizations and, where appropriate, with national organizations after consultation with the Member of the United Nations concerned.* (UN, 1945)

Several UN General Assembly and Economic and Social Council (ECOSOC) resolutions have deepened and formalized this relationship further in the past decades. The most important ones are the UN Resolution 1968/1296 and ECOSOC Resolution 1996/31,[3] establishing a solid legal framework for the partnership between civil society and the UN. Concrete examples of this growing degree of involvement of civil society can be found in development policies (Smillie et. al, 1999; Weiss, 1998), but also in the growing participation of civil society actors in world summits[4] (UN, 2001b).

By involving civil society, international – as well as national – institutions try to re-establish their legitimacy as operating in the interest of all and being democratically accountable, at least at the rhetorical level. The question is how this multi-stakeholder approach using participatory discourses materialises in a real life context where vested interests are at play, as well as processes of change. Before we come to this it remains important to (re-)articulate the notion of participation.

What is Participation?

Participation is an ideologically loaded and highly contested notion. For instance Pateman (1970: 1) remarks: *'the widespread use of the term [...] has tended to mean that any precise, meaningful content has almost disappeared; 'participation' is used to refer to a wide variety of different situations by different people'*. Different strategies have been developed to cope with this significatory diversity, most of which construct categorization systems. As the illustrations below will show, the element that supports the construction of these systems is the degree to which power is equally distributed among the participants. For this reason, the key concept of power will be addressed in a second part.

Constructing Participation as 'Real'

This widespread use (or the floating) of (the signifier) participation has firstly prompted the construction of categorising systems based on the combination of different concepts. In the context of the UNESCO debates about a 'New World Information and Communication Order' (NWICO)[5] the distinction between access and participation was introduced. While their definition of access stressed the availability of opportunities to choose relevant programs and to have a means of feedback, participation implied *'a higher level of public involvement [...] in the production process and also in the management and planning of communication systems.'* (Servaes, 1999: 85) Within communication studies, attempts have been made to introduce the notion of interaction as an intermediary layer between access and participation (Grevisse & Carpentier, 2004). From a policy studies perspective, complex typologies have been developed to tackle all variations in meaning – see for instance Arnstein's ladder of citizen participation (1969). More useful in this context is the OECD's (2001) three-stage model, which distinguishes information dissemination and consultation from active participation.

Other authors have aimed to construct hierarchically ordered systems of meaning, in which specific forms of participation are described as 'complete', 'real' and 'authentic', while other forms of participation are described as 'partial', 'fake' and 'pseudo'. An example of the introduction of the difference between complete and partial participation can be found in Pateman's (1970) book *Democratic Theory and Participation*. The two definitions of participation that she introduces are of 'partial' and 'full' participation. Partial participation is defined as: *'a process in which two or more parties influence each other in the making of decisions but the final power to decide rests with one party only'* (Pateman, 1970: 70 – our emphasis), while full participation is seen as *'a process where each individual member of a decision-making body has equal power to determine the outcome of decisions.'* (Pateman, 1970: 71 – our emphasis) Other terms have been used to construct a hierarchically ordered system within the definitions of participation on the basis of the real–unreal dichotomy. In the field of so-called 'political participation', for example, Verba (1961: 220–221) indicates the existence of 'pseudo-participation', in which the emphasis is not on creating a situation in which participation is possible, but on creating the feeling that participation is possible. An alternative name, which is among others used by Strauss (1998: 18), is 'manipulative participation'.[6] An example of an author working within the tradition of participatory communication who uses terms as 'genuine' and 'authentic' participation is Servaes. In *Communication for development* (1999) he writes that this 'real' form of participation has to be seen as participation '*[that] directly addresses power and its distribution in society. It touches the very core of power relationships.*' (Servaes, 1999: 198 – our emphasis) The concept of power is in other words again central to the definition of 'real' participation. White (1994: 17) also emphasises this central link between power and participation:

> *it appears that power and control are pivotal subconcepts which contribute to both understanding the diversity of expectations and anticipated out-comes of people's participation.* (our emphasis)

Power

If power is granted this crucial role in the definitional play, the need arises to elaborate further the meaning of the notion of power. In order to achieve this, we can make good use of the defining frameworks developed by Giddens and Foucault. Both authors stress that power relations are mobile and multidirectional. Moreover they both claim that their interpretations of power do not exclude domination or non-egalitarian distributions of power within existing structures. From a different perspective this implies that the level of participation, the degree to which decision-making power is equally distributed and the access to the resources of a certain system are constantly (re-)negotiated.

Both authors provide room for human agency: in his dialectics of control Giddens (1979: 91) distinguishes between the transformative capacity of power – treating power in terms of the conduct of agents, exercising their free will – on the one hand, and domination – treating power as a structural quality – on the other. This distinction allows us to isolate two components of power: transformation or generation (often seen as positive) on the one hand, and domination or restriction (often seen as negative) on the other. In his analytics of power, Foucault (1978: 95) also clearly states that power relations are intentional and based upon a diversity of strategies, thus granting subjects their agencies.

At the same time Foucault (1978: 95) emphasizes that power relations are also '*non-subjective*'. Power becomes anonymous, as the overall effect escapes the actor's will, calculation and intention: '*people know what they do; they frequently know why they do what they do; but what they don't know is what what they do does*' (Foucault quoted by Dreyfus and Rabinow (1983: 187)). Through the dialectics of control, different strategies of different actors produce specific (temporally) stable outcomes, which can be seen as the end result or overall effect of the negation between those strategies and actors. The emphasis on the overall effect that supersedes individual strategies (and agencies) allows Foucault to foreground the productive aspects of power and to claim that power is inherently neither positive nor negative (Hollway, 1984: 237). As generative/positive and restrictive/negative aspects of power both imply the production of knowledge, discourse and subjects, productivity should be considered the third component of power.[7]

Based on a Foucauldian perspective one last component is added to the model. Resistance to power is considered by Foucault to be an integral part of the exercise of power. (Kendall and Wickham, 1999: 50) Processes engaged in the management of voices and bodies, confessional and disciplinary technologies will take place, but they can and will be resisted. As Hunt and Wickham (1994: 83) argue:

Power and resistance are together the governance machine of society, but only in the sense that together they contribute to the truism that 'things never quite work', not in the conspiratorial sense that resistance serves to make power work perfectly.

Figure 1: Foucault's and Giddens' views on power combined

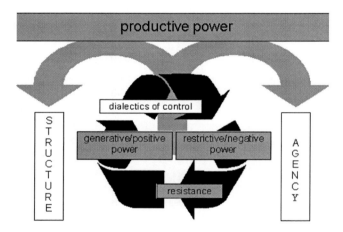

By using this model both the more localized and the more generalized power practices can be taken into account. This also allows us to bypass some of the problems that complicate the use of the notion of participation. Instead of almost unavoidably having to put an exclusive focus on the degree of structuralized participation, this theoretical framework emphasises the importance of localized and fluid (micro)power practices and strategies without ignoring the overall (political) structure. Firstly, this approach also allows stressing the importance of the outcome of this specific combination of generative and restrictive (or repressive) power mechanisms. The overall effect – the discourses, identities and definitions that were produced – will have their impact on future processes. By building on the analysis of the dialectics of control, we moreover argue here that the comparison of the generative and restrictive (or repressive) power mechanisms allows establishing the depth and quality of civil society participation in the WSIS process.

Following Foucault and Giddens we fully realise the existence of unequal power relations, and only use the notion of full participation as a democratic imaginary or utopia. This type of 'not-place' and 'never-to-be-place' provides this chapter with an ultimate anchoring point, which will always remain an empty place. Despite the impossibility of its actual realization in social praxis, its phantasmagoric realization serves as the breeding grounds for civil society's attempts oriented towards democratization. As the French writer Samuel Beckett of Irish decent once eloquently formulated it[8] *'Ever tried. Ever failed. Never mind. Try again. Fail better.'* In social practice we remain confronted with persistent power imbalances, but the social imaginary of full participation can be applied to legitimate (and understand) our plea for the maximization of generative and the minimization of restrictive power mechanisms.

In what follows this model of productive, generative and restrictive power mechanisms and the resistance they provoke, will be applied to the WSIS process. In order to do so a distinction will be made between the level of access/consultation to the process and the level of participation to the process, whereby the latter refers to the capacity to change or influence process-related outcomes.

Access and Participation in the WSIS and its Preparatory Process

In view of its longstanding partnership with NGOs the UN considered the involvement and participation of civil society in the World Summit on the Information Society (WSIS) to be paramount. UN Resolution 56/183 encouraged:

> *intergovernmental organizations, including international and regional institutions, non-governmental organizations, civil society and the private sector to contribute to, and actively participate in the intergovernmental preparatory process of the Summit and the Summit itself.* (UN, 2001a: 2)

In this regard, the Executive Secretariat of the WSIS created a Civil Society Division team that was given the task *'to facilitate the full participation of civil society in the preparatory process leading up to the Summit'* (our emphasis). The WSIS is also one of the first summits where ICTs are being used extensively to facilitate the interaction between the UN institutions and civil society actors, be they transnational, national or local. It is also the first world summit where civil society has been involved in the preparatory process from the very beginning. In many ways the WSIS was presented as a model for the new multi-stakeholder approach followed by the UN. As elaborated before, the notion of power is considered the crucial defining element in the discussion on participation. In this section we will analyse and compare both the generative and restrictive power mechanisms, as well as acts of resistance that are at play within (and also outside) the WSIS process. Moreover the overarching signifier participation will be split into two segments: access/consultation and participation. This distinction is important, as it allows highlighting the difference between being able to attend and observe the process (access), having one's opinions heard (consultation) and actually being able to influence the outcome of the process (participation).

In order to do so we will use several methodologies, which will allow us to get the broadest possible overview.[9] First of all, a quantitative data analysis of the attendance and accreditation lists provided by the ITU of the different PrepComs and the summit itself will allow us to assess access in detail. Despite the high degree of detail of these lists, using them also makes us dependent upon the registration process and its margins of error. Secondly, desk research of the WSIS-related websites, official documents/resolutions, as well as evaluation documents drawn up by key civil society actors will allow us to analyse civil society consultation and participation. Finally, in view of validating our research, a number of key persons[10] were invited to comment upon a draft version of this chapter.

Generative/Positive Power Mechanisms

In this first part, the first component of the theoretical model on participation and power is analysed. Here our analysis thus focuses on the generative aspects of civil society's role at the WSIS, both at the level of access and consultation, and at the level of participation.

Physical Access to the PrepComs and Summit

The WSIS is one of the first world summits where civil society was given extensive access to the preparatory process and to be present at the meetings.[11] In this context 'being present' refers to being able to access the meetings and being given limited speaking rights. As such, civil society is being recognized as a legitimate actor to be consulted on specific issues and to provide feedback allowing for a dialogue between civil society, state actors and the UN to take place.

During the summit and the build-up to it, the number of members from civil society, as well as civil society organizations (CSOs[12]) present, was quite high and grew steadily from PrepCom1 to the WSIS-03 in Geneva (cf. figure 2). There were 178 civil society members from 102 CSOs that attended the PrepCom1 meeting, which dealt primarily with procedural issues. About half of these organizations, however, did not attend PrepCom2 in Paris, which was held a year later. Nevertheless, at PrepCom2, 344 members from 176 CSOs were present. Here also some 70 organizations that attended PrepCom2 did not attend PrepCom3. At PrepCom 3, a decisive moment in the agenda-setting process and the drafting of the final declaration, the number of members from civil society increased to some 500 from 224 CSOs. The outflow from PrepCom3 to the WSIS was much lower, namely about 45 organizations. At the summit itself attendance of civil society rose to about 3200 members[13] from 453 CSOs.

Figure 2: Inflow, outflow and re-inflow of active CSOs[14] in the PrepComs and WSIS-03

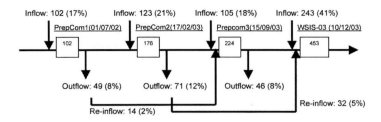

Figure 2 and table 1 also show that the re-inflow of CSOs did occur, although it remained rather limited. Fourteen CSOs attended PrepCom1, were absent at PrepCom2, but did attend PrepCom3 and 32 CSOs were not present at PrepCom1 and PrepCom3, but did attend PrepCom2 and the summit itself. Besides this, table 1 also shows that some 75% of the active CSOs were present at WSIS-03 in Geneva, which is quite high. The fact that some 25% of the CSOs that were active within the WSIS PrepCom process did not attend the WSIS itself in Geneva may have many reasons. It is, however, difficult to assess at this stage what these reasons were, but it could be that some organizations felt disillusioned with the process and/or did not have enough resources to remain actively involved. The data, however, also show that a big proportion of active CSOs only became involved at a late stage in the process. For some 40% of active CSO organizations the summit itself was the first time they were visibly involved in the process and only 7% of active CSOs attended all PrepComs as well as the summit itself. But, at the same time, the data reveal that very active organizations tended to remain involved throughout the process. Only 2% of active CSOs disengaged from the process, although they had attended PrepCom2 and 3 (presence at PrepCom1 is disregarded).

Table 1: Re-inflow & degree of attendance in PrepComs and WSIS 2003 in Geneva

PrepCom1	PrepCom2	PrepCom3	WSIS-03	#ORGs	%act-CSO	
+	+	+	+	40	7%	Present at all PrepComs and WSIS03
	+	+	-	12	2%	Present at PrepCom2 and 3, but not at WSIS03
-	-	+	+	74	12%	Not present at PrepCom1&2, but present at PrepCom3 & WSIS03
-	-	-	+	230	39%	Not present at PrepComs, but present at WSIS03
			*	453	76%	Present at WSIS03
			Total	595	100%	
Legend:	+: Present	-: Not present	*: Presence disregarding previous meetings			

The analysis of attendance in terms of the distribution over the different continents and the type of CSO, provides us with another angle (cf. fig. 3). From a generative point of view, it has to be noted that attendance from African CSOs in the WSIS process was quite high, accounting for almost 20% of active CSOs. Also noteworthy is the high attendance of academics in the WSIS process, about 15% of the active CSOs. Furthermore, it can also be asserted that attendance by local CSOs largely outweighs the presence of international and regional CSOs, more than 50% of active CSOs are locally based. This is a positive sign in terms of representing specific local contexts and concerns.

Figure 3: Regional distribution of CSOs by type of CSO[15]

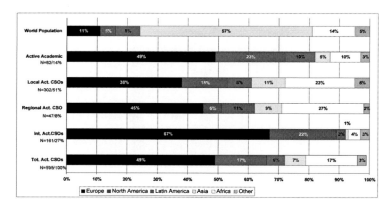

The ITU also held five regional meetings.[16] Attendance in these regional meetings was quite high. For example, more then 1700 participants[17] were present at the regional meeting in Bamako and as such it can be suggested that this helped in lowering the threshold for access to the WSIS process. Each regional meeting resulted in a declaration highlighting the demands and concerns of that particular region and also produced several documents.[18] In the case of the Bamako regional meeting, UNESCO, together with the Executive Secretariat, also organized a consultation round with African CSOs, which was subsequently reported in a document that can be found on the website of the Bamako meeting.

Virtual Access to the PrepComs and Summit

In addition to the access granted to CSO and the resulting 'offline' consultation rounds, the ITU Executive Secretariat developed an online platform specifically directed at involving, amongst others, civil society actors in the WSIS preparation process.

Accredited entities were encouraged to submit written contributions to the Executive Secretariat, who would then post them on the WSIS website and thus make them available for consultation. As such (accredited) CSOs could provide input for the summit declaration and the draft plan of action to be discussed in the preparatory process and to be voted by the member states in Geneva, mid-December 2003. As PrepCom1 dealt with procedural issues it is not surprising that especially during PrepCom2 and PrepCom3 many organizations contributed to the process (cf. table 2). During PrepCom2 the civil society organizations were very active, which amongst others is exemplified by the relatively high number of organizations (75/214=35%) that submitted a document vis-à-vis those organizations that were active during PrepCom2.[19] During PrepCom3 this percentage dropped slightly to 27%. By making the distinction between CSOs that introduced their own document and CSOs that co-signed documents with other organizations it is possible to make an assessment of the degree of networking. This is especially relevant for PrepCom2 where two-thirds of CSOs that submitted a document did this together with other organizations. It should be noted that during PrepCom2 negotiations started concerning the agenda and themes for the final declaration and that this explains the high degree of networking and documents being produced.

Table 2: Written contributions submitted to the preparatory process

	PrepCom1	PrepCom2	PrepCom3
#CSOs[20] that submitted their own document (a)	2	29	51
#CSOs that co-signed a document (b)	15	46	19
#CSOs with their own or co-signed document (c=a+b)	17	75	70
#CSOs with doc that did not attend PrepCom (d)	0	38	29
#CSOs attending PrepCom (e)	102	176	227
#CSOs active within PrepCom (f=d+e)	102	214	256
%CSO with doc of #CSOs active within Prepcom (g=c/f)	17%	35%	27%

Besides the written contributions, all the interventions made by heads of states, ministers, private sector representatives and civil society members during the plenary sessions of WSIS-03 in Geneva were recorded, webcasted, and have been archived on the ITU site for everyone to view and listen to. Also the press conferences were webcasted as well as archived. In general terms it can be asserted that the ITU has been very open in publishing contributions to the WSIS process and making them available for all to read, view, or listen. Upon demand of the Civil Society Caucus the critical alternative civil society declaration *Shaping Information Societies for Human Needs* (Civil Society Plenary, 2003) was also posted alongside the official declaration on the official WSIS website.[21]

The civil society caucuses also extensively used a number of mailing lists as a powerful tool to discuss issues and common strategies regarding the WSIS and shaping the agenda. In addition to these mailing lists to which civil society members could subscribe they also developed a virtual *'WSIS Civil Society Meeting Point'*[22] giving access to the mailing lists, addresses and those responsible for co-ordinating the different caucuses. This proved to be very successful and is still active in the run-up to the second phase of the WSIS in Tunis.

Participation

In this part we examine not so much the physical presence – the access to the process – nor the resulting consultation rounds but rather the rules that apply to structure the presence and its generative effects on the participation of civil society in the formal process. Besides this, there are also generative mechanisms at play on an informal level in terms of, for instance, networking.

The formal rules making the participation of CSO in world summits possible are based on the ECOSOC 1996/31 resolution passed by the 49th plenary meeting in July 1996. This resolution serves as a general guideline defining the consultative relationship between civil society and the UN. Part VII of the resolution deals specifically with what they called the *'participation of non-governmental organizations in international conferences convened by the UN and their preparatory process'*. Besides the conditions for accreditation, which relates more to access, civil society actors are given some rights in the ECOSOC resolution.

> *51. The non-governmental organizations accredited to the international conference may be given, in accordance with established United Nations practice and at the discretion of the chairperson and the consent of the body concerned, an opportunity to briefly address the preparatory committee and the conference in plenary meetings and their subsidiary bodies.*

> *52. Non-governmental organizations accredited to the conference may make written presentations during the preparatory process in the official languages of the United Nations as they deem appropriate. Those written presentations shall not be issued as official documents except in accordance with United Nations rules of procedure.* (ECOSOC, 1996/31)

The above-mentioned resolution is, however, a frame of reference and each summit can decide upon other modalities for participation going beyond 1996/31. The rules of procedure being adopted by each world summit define the nature of civil society involvement and govern the participation of civil society actors within the preparatory process, as well as the summit itself. Rule 55 of the WSIS rules of procedure, adopted during the 1st PrepCom (July 2002), relates to the participation of non-state actors, including CSOs.

> *Rule 55*
> *Representatives of non-governmental organizations, civil society and business sector entities*
> *1. Non-governmental organizations, civil society and business sector entities accredited to participate in the Committee may designate representatives to sit as observers at public meetings of the Preparatory Committee and its subcommittees.*
> *2. Upon the invitation of the presiding officer of the body concerned and subject to the approval of that body, such observers may make oral statements on questions in which they have special competence. If the number of requests to speak is too large, the non-governmental organizations, civil society and business sector entities shall be requested to form themselves into constituencies, such constituencies to speak through spokespersons.* (WSIS, 2002a)

Another document called *Arrangements for Participation*, jointly published with the rules of procedure, calls upon accredited NGOs and business-sector entities *'to actively participate in the intergovernmental preparatory process and the Summit as observers'* (WSIS, 2002a). Furthermore, it also encourages NGOs and business-sector entities to submit written contributions and pledges to post these on a website and to distribute the executive summaries to member-state representatives and other interested parties.

Besides the formal rules allowing CSO to be present and to present their points of view, there are also clear signs that summits also play an important role in terms of informal processes (maybe more so then formal) and network practices (Padovani, 2004a). Bridges (2004) refers to this when evaluating the WSIS process:

> *Simply by bringing so many stakeholder to the same place, WSIS helped stimulate partnerships. [...] Though this type of international collaboration is not reflected in the official paper trail, WSIS helped facilitate ground-level connection that will bring ICTs to a more prominent place on the world stage.*

Opening up the preparatory process and world summits to civil society, has generated its own dynamic in terms of informal contacts, mailing lists and lobbying efforts. Although the real impact of civil society on the formal level is qualified as rather low by many CSO representatives, most agree on the big success the summit was in terms of networking amongst civil society organizations and activists. This will not necessary show in the documents or the institutional level of analysis but has to be placed in a long-term perspective (Ó Siochrú, 2004b).

Summits such as the WSIS are also instrumental as learning experiences for civil society. In order to be taken seriously at a global level, civil society has to tackle criticisms by governments relating to a lack of representativity and the inability to speak with a *'co-ordinated voice'* (Kleinwächter, 2004: 1). By issuing an alternative declaration stating its own distinct positions, as well as by having some impact on the formal agenda, the civil society caucus has shown that civil society as an actor in processes of global governance is growing in its new role.

Lastly, some states have also been creative in order to allow civil society actors not only access to the process (as observers), but also enabled their participation in the process. Germany and Canada, for example, incorporated civil society representatives in their official 'state' delegations that attended the WSIS. As such they undermined the formal – and fairly strict – rules, and incorporated civil society within the formal structure of an official delegation.

Restrictive/Negative Power Mechanisms

In the second part of our analyses we return to the second component of the theoretical model on participation and power: the restrictive aspects of power. When analysing the restrictive or negative power mechanisms at the level of practices, the same distinction between physical access, virtual access and participation is made.

Physical Access

Although UN resolution 56/183 was quite ambitious in wanting to involve civil society, diplomatic pressures to limit the scope and extent of civil society involvement were also at play. Contrary to countries like Germany and Canada, many 'repressive' or at least authoritarian countries were not so keen on opening up a world summit to civil society. Governments like Pakistan and China made it very clear that they, and not the (often opposition) CSOs, represent their citizens (Hamelink, 2003; Toner, 2003). Besides the reluctance of some countries to involve civil society, other countries were not so keen on the WSIS as such for ideological and political reasons. When George Bush Jr. came to office, the US largely abandoned its digital divide discourses and policies developed by Clinton and Gore. Furthermore, the current US administration is, generally speaking, less interested in committing itself to international summits and agreements (cf. Kyoto Agreement).

There are, however, a number of other restrictive mechanisms that limit the CSOs' access to the process. In a discussion paper, *Civil Society Participation in the WSIS*, drafted by Seán Ó Siochrú and Bruce Girard (2002: 7–8) in order to prepare the WSIS process – a number of constraints to the access of civil society are enumerated. (1) Firstly, they identify a lack of structural funds and resources in order to allow civil society representatives to participate and attend the preparatory meetings and/or summit. (2) Secondly, this leads to a geographical imbalance. CSOs from poorer countries of the world are often *'unable to have their voice heard effectively'* and are increasingly dependent on big 'intermediary' civil society organizations to represent them and their constituencies. (3) Thirdly, the authors also refer to the poor presence of women.

The mechanisms Ó Siochrú and Girard describe result in processes of exclusion and the restriction of access of CSOs that find themselves in a less advantaged situation. But these differences are not only related to the more structural elements (such as the political-economic geography). Differences in access are also constructed on the basis of being categorized as part of civil society itself. The access of CSOs is firstly regulated through a system of accreditations, whereby the Executive Secretariat and the member states control the gate. Gaining entry trough the first gate is followed by a series of other forms of management (related to categorization, conflation, separation and surveillance) that further construct the difference between civil society and state actors and that limit the CSOs' abilities.

Excluding the Distant

Although the attendance of the African CSOs was deemed to be relatively high from a generative perspective, Western European civil society actors are still predominantly present. The reasons for this are of course complex and multiple. Our data for example suggests that almost all CSOs from Africa active within WSIS are quite young organizations (end of the 1990s, beginning of 2000). This confirms work on the recent wave of democratic reforms in Africa (Bratton & Van de Walle, 1997). Although difficult to prove within this research design it is conceivable that a lack of resources and experience in terms of global governance play a constraining role. This might explain the gap between the large proportion of African CSOs that showed an interest in the WSIS process and those who actually were able to attend. From all CSOs who have showed an interest in the WSIS process, some 40% came from Africa. Their share drops to 17% when only those CSOs that have been active within the WSIS process are taken into account (cf. fig. 4). Asia, on the other hand, is clearly under-represented. Human rights violations and the many rather authoritarian regimes in Asia could be one explanation for having a negative impact on civil society attendance from that region of the world. The dominance of Western languages, such as English, French and Spanish might also play a constraining role in this regard.

Figure 4: Regional distribution of active and non-active CSOs

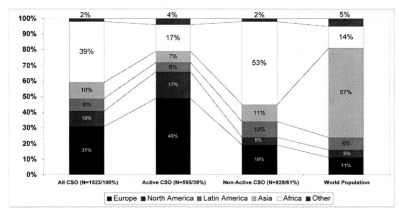

European – and to a lesser extent also North American – dominance also shows in the number of participants per organization[23] (cf. table 3). Most European and North American CSOs have three or two participants per CSOs, while most African CSOs present at WSIS-03, only have one participant. Latin America and Asia are in between with a median of two participants per participating CSOs. The fact that Geneva is one of the most expensive cities in Europe in terms of accommodation and cost of living and that travel costs from poorer regions in the world are generally speaking much higher might also explain why the number of participants from these regions is much lower.

Table 3: Average # of participants per civil society organization for WSIS-03 in Geneva

	Total	West Europe	East Europe	North America	Latin America	Southern SSAfrica	Arab World	Asia	Oceania	Unknown
# of CSO-participants	3205 (100%)	1977 (62%)	26 (1%)	599 (19%)	86 (3%)	204 (6%)	165 (5%)	138 (4%)	4 (<1%)	6 (<1%)
# of CSO	462 (100%)	208 (45%)	8 (2%)	85 (18%)	32 (7%)	54 (6%)	35 (5%)	33 (4%)	3 (<1%)	4 (<1%)
Average # Participants/CSO	6,9	9,5/7,1[24]	3,3	7,0	2,8	3,7/3[25]	4,7	4,1	1,0	-
Median # Participant/CSO	2	3	3	2	2	1	3	2	1	-

This also shows that despite the rhetorics of time–space implosions, 24h/7days-communication and increased mobility, major spatial constraints are still at play.

Management through Accreditation: The First Gate
As in each summit, access to the preparatory committees, as well as the summit itself, is dependent on getting an accreditation by the PrepCom of the summit. In essence there are two gatekeepers. The Executive Secretariat, in conjunction with the UN Non-governmental Liaison Service, evaluated applications and gave a

recommendation to the PrepCom who then took a decision. It is however not very clear what precise criteria were applied in this evaluation. In a document relating to the accreditation process drawn up before PrepCom1 it is stated very generally that:

> *The Executive Secretariat will review the relevance of the work of the applicants on the basis of their background and involvement in information society issues.* (WSIS, 2002b)

The Executive Secretariat will communicate its recommendations to the member states two weeks before the PrepCom. Member states can ask the Executive Secretariat for additional information and if they deem that not all conditions are met or that there is insufficient information, the PrepCom can defer its decision until its next meeting. It has to be remembered in this regard that civil society actors are only observers within the PrepCom and that it is the member states that decide (also with regard to accreditation). Furthermore, the provisions for appeal and the obligation for the Executive Secretariat to communicate the reasons for a negative recommendation to the concerned CSO, as foreseen in ECOSOC resolution 1996/31 (paragraph 46–47), is not mentioned at all in the WSIS arrangements for accreditation.

A notable example of an organization that was excluded from attending PrepCom3 and the summit was Reporters without Borders (Hudson, 2003). Reporters without Borders reacted by setting up a pirate radio station in order to protest against their exclusion and against a number of police activities during the summit (cf. 4.3). Before that the organization Human Rights in China was also excluded from the process without being given a reason why (HRIC, 2003). These two cases led to a bitter reaction from Meryem Marzouki, the co-ordinator of the WSIS human rights in the information society caucus:

> *A summit on the information society that allows the participation of governments that systematically censors medias and violates human rights but that doesn't allow the participation of some of the leading international groups defending those rights makes no sense.* (WSIS civil society media and human rights caucuses, 2003)

Management through Categorization, Conflation, Separation and Surveillance: The Second Gate

The ITU set up the Civil Society Division team in order to mediate between civil society on the one hand and ITU and organising committee on the other, but also to facilitate the involvement of civil society in the preparatory process. Although the signifier 'facilitation' might have a generative connotation, it also proved to be restrictive in its operation. Examples that could be found within the WSIS process are management through (1) categorization & conflation, (2) separation and (3) security, surveillance.

(1) Civil society subdivided itself in different caucuses and working groups.[26] This allowed for discussions and debates within civil society to be conducted in a more efficient and productive way. However, due to this categorization the Civil

Society Division team, set up by the ITU, was also able to assert itself as an interlocutor between civil society and states, instead of providing neutral administrative support.

Also noteworthy in this regard is the fact that local authorities[27] have been granted the status of civil society actors, while they are in fact state actors in stricto senso (Padovani, 2004b). Also business (related) actors such as the World Economic Forum or the International Chamber of Commerce were often referred to as civil society actors. This (intentional or unintentional) conflation of what constitutes civil society in fact also weakens its position.

(2) The second restrictive practice with regard to management of the process is the spatial and physical separation between official delegations and civil society participants. As such, space is also restricting at a micro-level and not only in terms of physical distance. Already during PrepCom1, where procedural issues were discussed, this proved to be a major disappointment for many civil society activists, as Alan Toner (2003: 10) asserts:

> *NGO participants discovered that while decisions on procedural form were to be discussed in the ITU building (where Pakistan and China were doing their utmost to have participation limited strictly to state-actors), they themselves were to be quarantined across the road where a programme of discussions had been scheduled for them by the Civil Society Directorate.*

This spatial separation between civil society actors and state actors also occurred to a lesser extent during the conference itself with separate restaurants, toilets, bars and sleeping arrangements for civil society participants and for state representatives. However, the summit venue itself was shared which (in theory) allowed for interaction between the different stakeholders.

(3) The third restrictive practice relates to security and how technology could (has) be(en) used to infringe the privacy of summit participants by processing information about their whereabouts during the summit (Hudson, 2003). Panganiban and Bendrath (2004) also condemned this in their evaluation:

> *The name badges produced for every summit participant at registration included a radio frequency identification (RFID) chip. The personal data of the participants, including the photograph, was stored on a central database, and the times when and where they left or entered the summit venue were also recorded. There was no privacy policy available, and nobody could or would tell us what happens to the data after the summit.*

We are not claiming here that privacy infringements have actually taken place, but it remains improper to issue participants with a name badge that has a tracking device in it, without telling them beforehand and without adopting a transparent privacy policy. When the researchers who discovered the presence of RFID chips in the name badges, asked the organisers what has been done with the data or for how long the data will be stored, they did not get any answer from the ITU (Hudson, 2003).

Some 2,000 military and 700 policemen also protected the summit. Security was very tight, with several 'checkpoints'. Security staff also screened leaflets and made a selection based on content, as reported by Sasha Costanza-Chock, a media activist active within the Campaign for Communication Rights (CRIS), quoted on dailysummit.net:

> *To inform the people, we do not have to go through metal detectors and checkpoints every 200 feet! We do not want to be in a space of controlled information, where they held me up yesterday and divided my papers and leaflets into two piles. The ones I could take in and the ones I couldn't.* (Constanza-Chock quoted in Obayu, 2003)

Virtual Access

ICTs are increasingly part of the global governance process as a facilitator for interaction and exchange of ideas between the UN and civil society actors, as well as amongst civil society actors in terms of networking and developing a common language (Cammaerts & Van Audenhove, 2003). Nevertheless, a tendency of overemphasising the role and impact of technology on this process can be perceived. The Internet is undoubtedly a powerful tool in many ways, but it is not the driving force of social change.

With regard to virtual access there are also a number of constraints that need to be taken into account. First of all there is the digital divide, precisely one of the most important issues raised by the WSIS, and the problems related to the unequal distribution of access to infrastructure, and to the lack of sufficient skills required to use the Internet and to process the overload of information. These problems necessitate a critical assessment of the access to the WSIS process through virtual means. As a study by O'Donnell (2003) showed, the digital divide is as real for citizens, as it is for CSOs. Especially organizations in poorer regions of the world have difficulties in terms of capabilities and access.

Besides this more general problem, the process of introducing a document also induces a number of restrictions. First of all, only accredited CSOs may submit documents. Secondly, the rules of procedure also stipulate that statements will be posted on the WSIS site, *'provided that a statement [...] is related to the work of the Preparatory Committee and is on a subject in which the non-governmental organization or the business sector entity has a special competence.'* (WSIS, 2002a: rule 57). This is quite vague and it is also not known if contributions have been refused or not. In any case, control resides with the Executive Secretariat. Also, the sheer number of documents to be found on the ITU site makes it very difficult (and painstakingly slow) to navigate through them. Besides this, very little effort has been put in synthesising the comments and contributions made by the different actors.

The WSIS-ITU sites also provided very little (or no) possibilities for interaction and discussion among citizens or civil society actors for that matter. The UNESCO forum for civil society was an exception to this, but although exposure was high, active engagement by civil society was rather low, making it an easy target for criticisms relating to representativity (cf. UNESCO, 2003).

Participation

Unfortunately, the WSIS process was not as open as projected in the official rhetoric and the opportunity to experiment with innovative co-decision mechanisms was not taken up. First of all, it has to be reiterated that the formal rules do not give civil society actors rights to vote in the preparatory process or the summit. The states still hold the negotiating role and the right to vote, as indicated in the ECOSOC 1996/31 resolution:

> 18. [...] the arrangements for consultation should not be such as to accord to non-governmental organizations the same rights of participation as are accorded to States not members of the Council and to the specialized agencies brought into relationship with the United Nations.
>
> [...]
>
> 50. In recognition of the intergovernmental nature of the conference and its preparatory process, active participation of non-governmental organizations therein, while welcome, does not entail a negotiating role. (ECOSOC, 1996/31)

Rule 55 in the formal rules of procedures for the WSIS, drawn up during PrepCom1, also clearly stated that civil society actors may designate representatives *'to sit as observers at **public meetings** of the preparatory committee and its subcommittees'* (WSIS, 2002a – our emphasis). In the ECOSOC resolution there is, however, no mention of limiting participation to public meetings. Ó Siochrú (2002: 1) pointed out that an earlier draft of the rules of procedure was much more open in this regard. It stated that representatives could sit as observers *'in the deliberations of the Preparatory Committee, and, as appropriate, any other subcommittee on questions within the scope of their activities'*. According to Ó Siochrú this *'stronger option'* was dropped after *'sustained opposition'* by some member states. Participation was reduced to the role of partial observer with the right to submit written contributions and with very restricted speaking rights.

Besides the formal rules restricting co-decision roles of civil society actors there are also more subtle restrictive practices at play. For example, the Civil Society Caucuses had set up an internal voting procedure, using its mailing lists, to select the representative of civil society to speak in the opening plenary. As such Lynne Muthoni Wanyeki from FEMNET (Kenya) and Carlos Afonso from RITS (Brazil) were suggested to the ITU by the Civil Society Caucus, but the ITU appointed Kicki Nordstrom, president of the World Blind Union instead.[28] Panganiban and Bendrath (2003) criticized this move by the ITU in their evaluation of the summit:

> We had selected our speakers in a fairly transparent and democratic manner before the summit. Then somebody in the ITU just took the list and arbitrarily picked and dropped people. We neither know who took this decision, nor why. But it denied civil society its right to choose who speaks on its behalf and brings its points across. This was especially clear in the opening ceremony. The selected speaker from the World Blind Union was nice, but had not participated actively in overall civil society discussions and therefore did not make our points. She even had been under pressure

from the ITU secretariat to include specific sentences in her speech. Oh, and by the way: This was even against the rules of procedure.

Those that did get the opportunity to voice the concerns and priorities of civil society actors found themselves with a very small audience (Sreberny, 2004: 195).

Resistance Against Restrictive Practices

Restrictive power mechanisms always induce and fuel different forms of resistance to these practices, which is the third component of the theoretical model on participation and power. Resistance manifested itself both within and outside the formal process. Here we will focus on resistance by the different stakeholders within the formal WSIS process, for more on resistance outside the formal process we can refer to the *'WSIS?We Seize!'* event organized by the Geneva03,[29] a collective of some 50 dissident CSOs. They organized five days of alternative events and actions. Besides this, one – rather marginal – demonstration was held against the WSIS, corporate control of information and in support of community media. It was organized on the last day of the WSIS (12/12/03) by an activist group called Collectif de résistance au SMSI (2003a; 2003b). They launched their appeal for action on the site of Indymedia–Switzerland. Only a mere 50 people showed up and were subsequently arrested or ordered to disperse by the more numerous security forces and police who were waiting for them. In effect, demonstrations were banned during the WSIS.

Resistance by Civil Society Actors

The ITU and the WSIS Executive Secretariat have supported and encouraged all actors within the WSIS process to stage side events in the fringes of or in conjunction with the formal process. In doing so other voices were generated and many governments, business actors and civil society organizations took this opportunity to organise such a side event, be it a symposium, a panel discussion, a forum, a seminar, a presentation or even concerts or exhibitions. In total 274 side events were set up.[30] Almost half of them were organized by civil society organizations. These can, of course, not all be labelled as resistance, but nevertheless many of these events used the forum of the WSIS to voice alternative discourses. For example, the *World Forum on Communication Rights* (11/12/03) organized by the campaign for communication rights (CRIS) attracted more then 600 participants. Another example was the *Community Media Forum* (12/12/03), organized by ALER, AMARC, Bread for All, CAMECO, Swiss Catholic Lenten Fund and the WSIS Community Media Caucus. Under the heading *WSIS?We Seize!* some 50 dissident CSOs also joined forces within the Geneva03 collective. They organized five days of alternative events and actions outside the formal setting of the WSIS.

Civil society did, of course, also resist many of the above-mentioned restrictive practices within the formal context of the WSIS. By denouncing them in the first place, but also in more subtle ways, for example by bending the rules. There is evidence of a struggle between states concerning the degree of involvement of civil society. Wolfgang Kleinwächter, an academic who has been from the very beginning a very close observer of the WSIS, gives an account of several instances where the minimalist Rule 55 was partially undermined and bent slightly to allow more participation by civil society.

> *The idea was, that governments, if they start negotiations on a certain paragraph, would interrupt formally the negotiations and invite observers to make a statement to the point. Such 'stop-and-go-negotiations' would de jure not change the character of inter-governmental negotiations, but could bring de facto innovative input and transparency to the process.* (Kleinwächter, 2004: 1)

The publication of an alternative declaration by the Civil Society Plenary (2003), is also clearly an act of resistance. In this regard it should be noted that also the Youth Caucus, the Swiss CS Contact Group, the Indigenous and the Disabilities Caucus issued alternative declarations, introducing different perspectives and thereby also voicing their dissent towards the official declaration agreed upon by governments at the WSIS in Geneva.

The maturation of civil society as an active and efficient actor in global or regional governance is of course also threatening as many CSOs challenge the dominant discourses and policies that many states and business actors put forward. Some argue that without civil society as an active and engaging 'observer' within the process, the declaration would have been 'even' less sensitive towards citizens' needs (Kleinwächter, 2004). In this regard it is disturbing to see indications that (some) states and business actors have disengaged from the summit (cf. next points). Others are less optimistic and see the official process as a stalemate, a consolidation of the market-driven approach of the Information Society notion and a rejection to consider alternative paradigms (Ó Siochrú, 2004a).

Resistance by Business Actors
During the PrepComs the number of written contributions by business sector actors, as well as their attendance could be qualified as rather low. In total some 125 companies, consultant firms or organizations representing corporate actors were active during the WSIS process. As table 4 shows business actors have not been very active in formulating their vision on the World information society within the (formal) PrepCom process. Their absence might be seen as a form of resistance towards possible changes.

It is, however, not unlikely that transnational corporations have other means available to get their views across, through lobbying governments directly or through the operations of so-called umbrella organizations. For example, many contributions from the business sector were produced by the Co-ordinating Committee of Business Interlocutors (CCBI) created by the International Chamber of Commerce (ICC) in order to '*mobilize and co-ordinate the involvement of the worldwide business community in the processes leading to and culminating in the Summit*' (ICC, 2003). Another important actor representing industry interests was ETNO, the European Telecom Network Operators.

Table 4: Participation of business sector actors and representatives

	PREPCOM1	PREPCOM2	PREPCOM3	WSIS-03
#Private Actors with Documents	4	7	6	-
#Private Actors Present	25	32	16	99
#Participants from Private Sector	35	61	30	513

Individual companies were much more reluctant to express themselves or be present at the meetings with their senior executives, let alone commit themselves to anything. Only 28 CEOs attended WSIS-03 in Geneva.[31] The senior management from Microsoft did not show up, nor many CEOs from major telecom operators or computer-hardware producers. The only 'big' industry players, within the information technology sector, who did send their CEO to the WSIS were Eutelsat (France), Nokia (Finland), Oracle (US), Fujitsu (Japan), Siemens (Germany) and Vodaphone (UK).

Resistance by States

Firstly, we can refer to the resistance of some states against increasing the role of civil society, which was voiced within the formal process and already mentioned above. Secondly, the turnout of heads of state and/or prime ministers at the WSIS-03 in Geneva was also quite meagre. As James Cowling (2003) asserts:

> *the combination of heads of state (many from the developing world) and lesser political figures from the rich countries was revealing: it was hard to avoid the impression that the latter took WSIS less seriously than the former.*

This also shows in our data. The US delegation, for example, was as big as that of Gabon (66 delegates). From the 176 states present at the WSIS-03, only 42 countries did send their (vice-)president or prime minister. Western heads of states were almost totally absent. Only Switzerland, being the host, France, Austria and Ireland were represented by their head of state or prime minister. Moreover, table 5 shows that contrary to Western European or North American reluctance, African countries, as well as the Eastern European countries with their emerging capitalist economies were very keen on sending their head of state to the WSIS. The number of Asian countries represented by a head of state or prime minister was also relatively high compared to the number of Asian CSOs that attended the WSIS.

Table 5: Number of countries that were represented by (vice-)president and/or prime minister subdivided by region

Regions:	#Countries	%	Regions:	#Countries	%
Western Europe	4	10%	Asia	5	12%
Eastern Europe	8	19%	Oceania	1	2%
North America	0	0%	Middle East	3	7%
Middle America	2	5%	North Africa	3	7%
Caribbean	0	0%	Sub-Sahara Africa	13	31%
South America	0	0%	Southern Africa	3	7%
TOTAL:			42 Countries (100%)		

Finally, resistance by states did not only show in the low attendance by heads of state, but also in comments and statements relating to the official declaration. David Gross, the US ambassador who represented the US administration in the preparatory process, was quoted on Daily Summit as stating: '*These are important documents, although they are not legally binding, [...] they are important expressions of political will.*' Furthermore, he reduced the scope of WSIS to technological issues and framed it as a political summit: '*It would be incorrect to see a political summit as a way to decide technological issues*' (Gross quoted in Malvern, 2003). The highest US representative at the summit, Bush's senior science and technical advisor, John Marburger III, emphasized the need for '*supporting technological innovation*', but did not mention the digital divide once during his plenary speech (Swift, 2003). This is problematic as these comments and statements undermined the whole effort of the summit and devaluated the reached consensus as formulated in the final declaration.

Resistance by states thus takes two contradictory stances. On the one hand by asserting that the WSIS deals with non-political, technological and economical matters, which implies that from a liberal perspective the state(s) should not intervene. On the other hand it is stated that the WSIS is 'not political enough', whereby the political is defined in a minimalist state-centred way, excluding civil society. From both perspectives civil society's role is discredited. The former interpretation excludes civil society, as the market is supposed to regulate itself and the latter interpretation excludes civil society because it is considered 'not-representative', and thus not politically legitimate.

Conclusion

The process of the WSIS, seen as a dialectics of control where generative, restrictive and resisting power mechanisms are at play, has 'produced' a series of outcomes. Following our Foucauldian perspective (which resulted in the inclusion of the fourth component in our theoretical model: production) these outcomes are the result of the unique combination of strategies and power games of all actors involved, without remaining blind for their embeddedness in clearly unequal power structures. Next to the more material output, such as the documents, the summit has also produced (new or perpetuated) inter-actor relationships, patterns of behaviour and discourses on participation.

It goes without saying that access to the WSIS was high and facilitated by several generative practices. To a lesser extent this can also be asserted about the consultation of civil society, by letting CSOs voice their concerns (mainly through written contributions). Also, if we put the WSIS in an historical framework and compare it with other summits, substantial advances were made (Selian, 2004; Raboy, 2004). At the same time, however, we have to conclude that the ambitious rhetoric of 'full' participation has not materialized. Even the partial participation (using Pateman's vocabulary) of civil society at the summit remains problematic, as the alternative declaration and the frustration among many civil society actors show. At most we can speak of a consultative process. By extensively using the notion of (full) participation, as well as the notion of '*citizenship in the information age*' (EU, 2002: 12), when, in fact, consultation and dialogue is meant, international organizations are on the one hand giving civil society a voice (generative power) but limit and restrain at the same time the impact of civil society (restrictive power). In contrast to this implicitly reductionist notion of participation being used by international organizations, we prefer to maintain a clear (as possible) distinction between access, interaction, consultation and participation, all embedded within the constant need to maximise the equalization of power relations (without ever reaching this 'ideal power situation', to paraphrase Habermas). This concurs with a more realist-step-by-step-approach towards making global governance processes more democratic.

From that perspective it can be concluded that summit negotiations aiming to reach a more globalized consensus are changing and shifting slowly towards increased – albeit informal – presence and consultation of civil society 'observers' within the (preparatory) processes of world summits. By fully taking up the opportunities given within the formal framework, by constantly moving the signposts of restrictions, by bending the rules, and by (at least more or less) speaking with a co-ordinated voice, the civil society caucus of the WSIS has asserted itself as a more mature partner in global governance. In this regard it is important to stress that the real outcome for civil society was not so much the formal process, on which it had a rather limited impact (Fücks, 2003; Dany, 2004), but the informal process of networking and mediation within civil society (Padovani, 2004a). It is especially here, but also in other ways, that the productive nature of the complex interplay between generative, restrictive and resistance practices shows itself most clearly.

However, if the rhetorics of increased participation (until now happily detached from its more radical meaning by international organizations and their member states) are to be fulfilled in the future, these international institutions should, amongst other actions, review the formal and legal rules that structure civil society 'participation', allowing for more equity, interaction and especially more moments of co-decisionship. Even more importantly, an in-depth reflection and consensus-building is required on the articulation of new definitions of two key components of democracy: participation and representation. These new definitions imply the broadening of participation beyond the limits of consultation, and the broadening of representation beyond the borders of political legitimacy through popular vote. Civil society from its part cannot ignore the learning opportunities offered by the experiences of the WSIS and needs to produce *'a new quality of balanced and substantial "positions" and "negotiable language"'*, thereby *'challenging governments'* to give substantial answers, making the whole process more transparent (Kleinwächter, 2004: 2). Furthermore, if the rhetoric of a global 'bottom-up' policy process is to be considered genuine, the civil society caucus, as well as international institutions, need to devise strategies in order to include more CSOs from Asia and Africa. It remains to be seen, however, if the second phase of the WSIS, which will take place in Tunis in 2005, will allow for more of this type of 'real' civil society participation in the new preparatory processes and the 2005 summit. Many civil society activists have already voiced their concerns on whether they will be allowed to participate freely and exert their right of free speech (cf. Civil Society Plenary, 2003: 21).

Finally, the observed disengagement and disinterest by some Western states and some business actors, partly resisting to the increasing number and influence of the dissident and critical voices emanated by civil society and to the demand of many developing countries for a digital fund, is a potential weakness in the multi-stakeholder approach. The realization of any action plan will require financial, as well as legal and political efforts from states (especially Western states), but also from the private sector since budgetary constraints limit (to some extent) the possibilities of states to act. If this trend continues, a clear danger arises that the WSIS as well as other summits (despite the UN's efforts to rethink civil society participation[32]) will end in an NWICO/UNESCO scenario of very ambitious goals and critical assessments, but no political – nor economical – will to actually turn even the watered-down declaration into a political reality.

Notes

1   This refers to a multi-centred world system where states are no longer the sole actors or stakeholders, but international organizations, business and civil society also play their role in global or regional governance. For more on this see Rosenau (1990); Hemmati (2002).

2   This analysis focuses on WSIS-03, the first phase of the WSIS; in 2005 the second phase will be held in Tunis.

3   The ECOSOC resolution is a review of the UN resolution.

4   At the Conference on Environment and Development in Rio de Janeiro (Brazil, 3–14/06/1992) some 2,400 representatives of non-governmental organizations (NGOs) were present and about 1,400 NGOs were accredited, 17,000 people attended the parallel NGO Forum. At the 4th World Conference on Women in Beijing (China, 4–15/09/1995) 5,000 representatives from civil society were present, some 2,100 civil society organizations were accredited and about 30,000 individuals participated in the independent NGO forum (Stakeholder Forum, 2002). At the World Conference against Racism, Racial Discrimination, Xenophobia and Related Intolerance in Durban (South Africa, 31/08–07/09/2001) some 1,300 NGOs were accredited (WCAR, 2001). Alongside the official conference an NGO forum was held in which 8,000 CS representatives from almost 3,000 CSOs participated (UN, 2001c: annex-v).

5   Or New International Information Order (NIIO).

6   The well-known rhyme, which according to myth appeared sometime around the beginning of the seventies on a Paris wall, also takes advantage of this dichotomy between 'real' and 'fake' participation: '*Je participe, tu participes, il participe, nous participons, vous participez, ils profitent.*' (Verba & Nie, 1987: 0).

7   Not all authors agree upon the distinction between the Foucauldian concept of productive power and the Giddean concept of generative power. We here follow the interpretation Torfing (1999: 165) proposes: '*Foucault aims to escape the choice between "power over" and "power to" by claiming that power is neither an empowerment, potentiality or capacity [generative power], nor a relation of domination [repressive power].*'

8   In order not to do history too much injustice: Samuel Beckett wrote these often quoted sentences in relation to art and not democracy unrealized.

9   This objective legitimises the use of both qualitative and quantitative methodologies. The theoretical framework used in this paper allows us to avoid any post-positivistic tendencies in the use of quantitative methods.

10 We would like to thank Robin Mansell, Cees Hamelink, Claudia Padovani, Seán Ó Siochrú, as well as the OII WSIS seminar participants and two anonymous reviewers for their very useful comments.

11 Three preparatory meetings or PrepComs were held in Geneva (PrepCom1, 01–05/07/02; PrepCom2, 17–28/02/03; PrepCom3, 15–26/09/03, 10–14/11/03 & 05–09/12/03) and one intersessional meeting in Paris (15–18 July 2003).

12 When speaking of Civil Society Organizations (CSOs) we mean those organizations that are independent from market and state, as such we adopt a Gramscian approach to the civil society notion (Cohen & Arato, 1990: ix). Thus, in the data-file we filtered out business actors and local authorities that were included in the lists of attending civil society organizations.

13 On a total of 10,808 participants.

14 By 'Active CSOs' we mean organizations that have been actively involved in the WSIS process, by participating in the PrepCom-meetings, the summit itself, and/or by submitting a document to the ITU-WSIS website. We do however acknowledge that being present at meetings or summit does not per se mean that organizations were very active in the process itself.

15 Other relates to Middle East, Oceania & Unknown (the same applies for figure 3).

16 Bamako (Mali, 5–30 May 2002), Bucharest (Romania, 7–9 November 2002), Tokyo (Japan, 13–15 January 2003), Bávaro (Dominican Republic, 29–31 January 2003), and Beirut (Lebanon, 4–6 February 2003).

17 Refers to the total number of participants. It was, however, not possible to distinguish between CSO and official (state) representatives.

18 For Bamako, see http://www.wsis2005.org/bamako2002/documents.html (last accessed 26/01/2005).

19 Again, by 'being active' we mean submitting a document and/or attending the PrepCom-meeting.

20 Data was collected from the ITU-WSIS website and is dependent on correct registration of attendance.

21 http://www.itu.int/wsis/

22 http://www.wsis-cs.org/

23 Due to single organizations with a very high number of participants, a correction was made for Europe and for Africa.

24 World Electronic Media Forum, based in Switzerland, does skew the results for Europe considerably, as they had 507 participants at the WSIS2003. For the average number of participants we made the calculations with WEMF included and excluded.

25 APC, based in South Africa, does skew the results of Africa as they had 47 participants to WSIS2003. For the average number of participants we made the calculations with APC included and excluded.

26 See http://www.wsis-cs.org/ for a full list (last accessed 26/01/2005).

27 As stated before we decided to disregard the local authorities in our data on CSOs.

28 For a full list of the CS speakers see URL: http://mail.fsfeurope.org/pipermail/wsis-euc/2003-December/000157.html (last accessed 26/01/2005).

29 For more on this see URL: http://www.geneva03.org/ (last accessed 26/01/2005).

30 For a full list see URL: http://www.wsis-online.net/event/events-list?showall=t (last accessed 26/01/2005).

31 For a full list see: http://businessatwsis.net/realindex.php (last accessed 26/01/2005).

32 cf. Secretary-General's Panel of Eminent Persons on Civil Society and UN Relationships, URL: http://www.un.org/reform/panel.htm (last accessed 26/01/2005).

References

Arnstein, S. R. 1969. 'A Ladder of Citizen Participation', *Journal of the American Institute of Planners* (35): 216–224.

Bauböck, B. 1994. *Transnational Citizenship*. Aldershot: Edward Elgar.

Beck, U. 1996. 'World Risk Society as Cosmopolitan Society?', *Theory, Culture and Society* 13(4): 1–32.

Boli, J., Thomas, G. M. 1997. 'World Culture in the World Polity: A century of international non-governmental organizations', *American Sociological Review* (62): 171–190.

Bratton, M., Van de Walle, N. 1997. *Democratic Experiments in Africa. Regime transitions in comparative perspective*. Cambridge: Cambridge University Press.

Bridges, M. 2004. 'WSIS – Conference Hype or Lasting Change?', *Berkman Briefings*, Harvard Law School, accessed 26/01/2005, http://cyber.law.harvard.edu/briefings/WSIS.

Cammaerts, B., Van Audenhove, L. 2003. *ICT-Usage among Transnational Social Movements in the Networked Society: to organise, to mediate & to influence*, EMTEL2 Key Deliverable. Amsterdam-Delft: UvA-TNO.

Civil Society Plenary. 2003. *Civil Society Declaration: Shaping Information Societies for Human Needs*, WSIS Civil Society Plenary, 8th of December, accessed 26/01/2005, http://www.itu.int/wsis/.

Cohen, J. L., Arato, A. 1992. *Civil Society and Political Theory*. Massachusetts: MIT Press.

Cowling, J. 2003. 'The internet's future in an aircraft hangar', 19/12, OpenDemocracy, accessed 26/01/2005, http://www.opendemocracy.org/debates/issue-8-10.jsp.

Dany, C. 2004. 'Civil Society and the Preparations for the WSIS 2003: Did input lead to influence?', accessed 26/01/2005, http://www.worldsummit2003.de/en/web/615.htm.

Dreyfus, H. L., Rabinow, P. 1983. *Michel Foucault: beyond structuralism and hermeneutics*. Chicago, Ill.: University of Chicago Press.

Economic and Social Council (ECOSOC). 1996. *Consultative relationship between the United Nations and non-governmental organizations*, Resolution 1996/31, adopted at 49th plenary meeting, 25/07, accessed 26/01/2005, http://www.un.org/documents/ecosoc/res/1996/eres1996-31.htm.

French, Marilyn. 2003. *From Eve to Dawn: A History of Women*. Toronto, Ont.: McArthur & Company.

Foucault, M. 1978. *History of Sexuality, part 1: an introduction*. New York: Pantheon.

Fücks, R. 2003. 'Preface', in Heinrich Böll Foundation (Ed.), *Visions in Process*. Berlin: Heinrich Böll Foundation/Heinrich-Böll-Stiftung, accessed 26/01/2005, http://www.worldsummit2003.de/download_de/Vision_in_process.pdf.

Geary, D. (ed.) 1989. *Labour and Socialist Movements in Europe Before 1914*. Oxford: Berg Publishers.

Giddens, A. 1979. *Central problems in social theory: action, structure and contradiction in social analysis*. London: Macmillan Press.

Grevisse, B., Carpentier, N. 2004. *Des Médias qui font bouger. 22 expériences journalistiques favorisant la participation citoyenne*. Brussel: Koning Boudewijn Stichting.

Hamelink, C. 2003. 'The Right to Communicate in Theory and Practice: A Test for the World Summit on the Information Society', Montreal, 13 November, 2003 Spry Memorial Lecture, Vancouver, 17 November 2003, accessed 26/01/2005, http://www.com.umontreal.ca/spry/spry-ch-lec.html.

Hauben, M. F. 1995. *The Netizens and Community Networks*, accessed 26/01/2005, http://www.columbia.edu/~hauben/text/WhatIsNetizen.html.

Held, D. 1995. *Democracy and the Global Order: From the Modern State to Cosmopolitan Governance*. Oxford: Polity Press.

Held, D. 1997. 'Democracy and Globalization', *Global Governance* 3(3): 251–267.

Held, D., McGrew, A., Goldblatt, D., Perraton, J. 1999. *Global Transformations: Politics, Economics and Culture*. Cambridge: Polity.

Hemmati, M. 2002. *Multi-Stakeholder Processes for Governance and Sustainability: Beyond Deadlock and Conflict*. London: Earthscan.

Hollway, W. 1984. 'Gender difference and the production of subjectivity', pp. 227–263 in J. Henriques; Hollway Wendy; Urwin Cathy; Venn Couze; Walkerdine Valerie (eds.) *Changing the subject. Psychology, social regulation and subjectivity*. London and New York: Methuen.

Human Rights in China (HRIC). 2003. 'HRIC Excluded From World Summit On the Information Society', Press Release, Human Rights In China, September 18, accessed 26/01/2005, http://iso.hrichina.org/public/contents/11607.

Hudson, A. 2003. 'Bug devices track officials at summit', 14/12, *The Washington Times*, accessed 26/01/2005, http://www.washingtontimes.com/functions/print.php?StoryID=20031214-011754-1280r.

Hunt, A., Wickham, G. 1994. *Foucault and law: towards a sociology of law as governance*. London: Pluto.

Hutchings, K., Dannreuther, R. (eds.) 1999. *Cosmopolitan Citizenship*. London: Macmillan Press.

International Chamber of Commerce (ICC). 2003. *Business input for World Summit on the Information Society*, International Chamber of Commerce, accessed 26/01/2005, http://www.iccwbo.org/home/e_business/wsis.asp.

Kendall, G., Wickham, G. 1999. *Using Foucault's Methods*. London: Sage.

Kleinwächter, W. 2004. 'WSIS: a New Diplomacy? Multi stakeholder approach and Bottom-up Policy in Global ICT Governance', evaluation paper.

Malvern, J. 2003. 'Status quo rocks', *Daily Summit*, 10/12, accessed 26/01/2005, http://www.dailysummit.net/english/archives/2003/12/10/status_quo_rocks.asp.

McBride, S. 1980. *Many voices, one world. Report by the international commission for the study of communication problems*. Paris & London: Unesco & Kogan Page.

Ó Siochrú, S., Girard, B. 2002. *Civil Society Participation in the WSIS: Issues and Principles*, Discussion Paper for Working Group 1 on Civil Society Participation in the WSIS.

Ó Siochrú, S. 2002. *Brief Analysis of Rules of Procedure and Accreditation*, CRIS-Info.

Ó Siochrú, S. 2004a. 'Will the Real WSIS Please Stand-up? The Historic Encounter of the "Information Society" and the "Communication Society"', *Gazette, the International Journal for Communication Studies* 66(3/4): 203–224.

O'Siochru, S. 2004b. 'Failure and Success at the WSIS: Civil Society's next moves', *UNRISD News* No. 26, spring/summer 2004, accessed 26/01/2005, http://www.worldsummit2003.de/en/web/599.htm.

Obayu, O. 2003. 'Our Space is Open', *The Daily Summit*, accessed 26/01/2005, http://www.dailysummit.net/english/archives/2003/12/11/our_space_is_open.asp.

O'Donnell, S. 2003. 'Civil Society Organizations and an inclusive Information Society in Europe', pp.77–96 in B. Cammaerts, et al. (eds.) *Beyond the Digital Divide: reducing exclusions, Fostering Inclusion*. Brussels: VUBpress.

Organization for Economic Cooperation and Development (OECD). 2001. *Citizens as Partners: Information, Consultation and Public Participation in Policy-Making*. PUMA: OECD.

Padovani, C. (ed.) 2004a. 'The World Summit on the Information Society. Setting the Communication agenda for the 21st century?', *Gazette, The international journal of communication*, special issue on WSIS, 66 (3–4): 187–191.

Padovani, C. 2004b. 'From Lyon to Geneva. What role for local authorities in the WSIS multi stakeholder approach?', Memo for the special issue of the Journal of Information Technologies and International Development, 'The World Summit in Reflection: a deliberative dialogue on WSIS'.

Panganiban, R., Bendrath, R. 2003. 'How was the Summit?', 16 December, the Heinrich-Böll-Foundation, Geneva/Berlin, accessed 26/01/2005, http://www.worldsummit2003.de/en/web/577.htm.

Pateman, C. 1970. *Participation and democratic theory*. Cambridge: Cambridge University Press.

Patomäki, H. 2003. 'Problems of Democratizing Global Governance: Time, Space and the Emancipatory Process', *European Journal of International Relations* 9(3): 347–376.

Rosenau, J. 1990. *Turbulance in World Politics, a Theory of Change and Continuity*. Princeton: Princeton University Press.

Sassen, S. 2002. 'The Repositioning of Citizenship: Emergent Subjects and Spaces for Politics', *Journal of Sociology* 46: 4–25.

Sassen, S. 1999. *Globalisering, over mobiliteit van geld, mensen en informatie*. Amsterdam: Van Gennep.

Saward, M. 2000. 'A Critique of Held', pp. 21–46 in B. Holden (ed.) *Global Democracy: Key Debates*. London: Routledge.

Schild, J. 2001. 'National vs European identities? French and Germans in the European multi-level system', *Journal of Common market studies* 39 (2): 331–351.

Servaes, J. 1999. *Communication for development. One world, multiple cultures*. Cresskill, New Jersey: Hampton Press.

Smillie, I., Helmich, H, German, T., Randel, J. (eds.) 1999. *Stakeholder: government – NGO partnerships for international development*. Paris & London: OECD & Earthscan Publications Ltd.

Sreberny, A. 2004. 'WSIS: Articulating Information at the Summit', *Gazette, the International Journal for Communication Studies* 66(3–4): 63–86.

Stakeholder Forum. 2002. UN Conferences focus, accessed 26/01/2005, http://www.earthsummit2002.org/roadmap/conf.htm.

Strauss, G. 1998. 'An overview', pp. 8–39 in F. Heller, E. Pusic, G. Strauss, B. Wilpert (eds.) *Organizational Participation: Myth and Reality*. New York: Oxford University Press.

Swift, C. 2003. 'Nothing Much to Say', 13/12, *The Daily Summit*, accessed 26/01/2005, http://www.dailysummit.net/english/archives/2003/12/11/nothing_much_to_say.asp.

Toner, A. 2003. 'Dissembly Language: Unzipping the World Summit on the Information Society', Autonomedia and New York's Information Law Institute, *Metamute*, M26: 4.07.03, accessed 26/01/2005, http://www.sindominio.net/metabolik/textos/ginebra.pdf.

Torfing, J. 1999. *New theories of discourse*. Laclau, Mouffe and Žižek. Oxford: Blackwell.

Tucker, K. H. 1998. *Anthony Giddens and modern social theory*. London: Sage.

United Nations Economic and Social Council. 2002. 'Implementing Agenda 21', *Report of the Secretary-General*, CN.17/2002/PC.2/7.

United Nations (UN). 1945. *Charter of the United Nations*, UN: San Francisco, accessed 26/01/2005, http://www.un.org/aboutun/charter/index.html.

UN. 2001a. *World Summit on the Information Society*, Resolution 56/183, adopted by the General Assembly, 21/12, UN: New York, accessed 26/01/2005, http://www.itu.int/wsis/docs/background/resolutions/56_183_unga_2002.pdf.

UN. 2001b. *Reference document on the participation of civil society in United Nations conferences and special sessions of the General Assembly during the 1990s*, Office of the President of the Millennium Assembly, 55th session of the United Nations General Assembly, accessed 26/01/2005, http://www.un.org/ga/president/55/speech/civilsociety.htm.

UN. 2001c. *Report of the World Conference against Racism, Racial Discrimination, Xenophobia and Related Intolerance*, UN: Geneva, accessed 26/01/2005, http://www.unhchr.ch/huridocda/huridoca.nsf/(Symbol)/A.Conf.189.12.En?Opendocument.

United Nations Educational, Scientific, and Cultural Organization (UNESCO). 2003. *Final Report NGOs and Civil society online discussion forum*, accessed 26/01/2005, http://portal.unesco.org/ci/ev.php?URL_ID=7820&URL_DO=DO_TOPIC&URL_SECTION= 201&reload=1081194315.

Urry, J. 2003. *Global Complexity*. Cambridge: Polity.

Van Steenbergen, B. 1994. 'Towards a global ecological citizen', pp 141–152 in Van Steenbergen, B. (ed.) *The Condition of Citizenship*. London: Sage.

Verba, S. 1961. *Small groups and political behaviour*. Princeton: Princeton University Press.

Verba, S., Nie, N. 1987. *Participation in America: Political Democracy & Social Equality*. Chicago: University of Chicago Press.

Vertovec, S., Cohen, R. (eds.) 2002. *Conceiving Cosmopolitanism: Theory, Context and Practice*. Oxford: Oxford University Press.

World Conference Against Racism (WCAR). 2001. 'List of NGOs not in consultative status with ECOSOC, World Conference against Racism, Racial Discrimination, Xenophobia and Related Intolerance', UN: Durban, accessed 26/01/2005, http://www.unhchr.ch/html/racism/05-ngolist.html.

Weiss, T.G. 1998. *Beyond UN subcontracting: task-sharing with regional security arrangements and service-providing NGOs*. London: Macmillan.

White, S. 1994. *Participatory communication: working for change and development*. Beverly Hills: Sage.

World Summit on the Information Society (WSIS) civil society media and human rights caucuses. 2003. 'Exclusion of Reporters sans Frontieres from the World Summit on the Information Society', Communique, accessed 26/01/2005, http://amsterdam.nettime.org/Lists-Archives/nettime-l-0309/msg00101.html.

WSIS. 2002a. *Report by the Chairman of Subcommittee 1 on Rules of Procedure*, WSIS/PC-1/DOC/0009, accessed 26/01/2005, http://www.itu.int/dms_pub/itu-s/md/02/wsispc1/doc/S02-WSISPC1-DOC-0009!!MSW-E.doc.

WSIS. 2002b. *Arrangements for accreditation*, adopted at the first session of the Preparatory

Committee, Geneva, 1–5 July 2002, accessed 26/01/2005, http://www.itu.int/wsis/documents/
doc_single-en-17.asp.

Zacher, M. 1992. 'The Decaying Pillars of the Westphalian Temple', pp. 58–101 in J. Rosenau
(ed.), *Governance without Government, Order & Change in World Politics*. Cambridge:
Cambridge University Press.

2: Communication Governance and the Role of Civil Society: Reflections on Participation and the Changing Scope of Political Action

CLAUDIA PADOVANI & ARJUNA TUZZI

Positioning 'Civil Society' on the Global Scene

Literature on global change is expanding, focusing on the multidimensionality of processes, the extension and deepening of social relations in different sectors: economy, politics, culture (Featherstone, 1990; Rosenau, 1992; Held et al., 1999). In this context a '*shifting in the location of authority*' is a crucial change (Rosenau, 1999); a shifting to spaces, where decisions are made, that are more and more distant from the people, making citizens perceive political processes as distant and '*opaque*' (Neveu, 2000).

Global change is therefore producing actions and reactions on the side of 'civil society organizations' that invite us to re-open a discussion on the practice of citizenship in a contemporary world and on the future of democracy, within states as well as beyond them. Civil society is an emerging actor in international politics (Baylies and Smith, 1999; Arts, 2003). Civil society can be considered '*an answer to war*' (Kaldor, 2003). It is being invited to '*actively participate in intergovernmental political processes*' (UN GA Res. 56/183, December 2001).

Dealing with civil society is a priority issue for the United Nations, as demonstrated by the ad hoc panel of independent experts set up by UN Secretary General Kofi Annan in July 2003, and chaired by former President of Brazil Fernando Henrique Cardoso, which issued its final report in June 2004.[1] In a very different sense, this is a priority also for the United States administration and its business partners, who have set up an 'NGOwatch' programme to monitor the growing number of lobbying NGOs, which is perceived as a threat to the sovereignty of constitutional democracies (Niggli, 2003).

In this context, a number of problematic questions arise: what conceptions of civil society underline these developments? How should we think of civil society as an actor in global politics? How does it get organized? And ultimately: what kind of power does/can it exercise?

Trans-national forms of organization 'from below' have a long history (Keck & Sikkink, 1998; Kaldor, 2003): in different fields, such as environment, human rights, gender, development and peace, relations have developed over time amongst social movements and grass-roots organizations. It is important to notice that the use of new communication technologies and the Internet nowadays sustain such relations. At the same time, they are strengthened in different 'spaces of place': occasions for physical meetings such as the World Conferences convened by the UN; protest events organized on the occasion of high-level political summits as in Seattle, Genoa or Cancun; or autonomous gatherings that set the landscape and agenda for a 'globalization from below' autonomously from official events, as in the case of the World Social Forum.

We should consider the political meaning of not-so-visible and yet profound transformations, such as the organizational and communicative competences developed within civil society networks (Keck & Sikkink, 1999; Smith et al., 1997) and their multi-level *modus operandi*. These are two developments that seem to be both a resource for and the result of a growing awareness of their political relevance. Moreover, these actors are in some cases critically self-reflecting, and developing a discourse on the role they should play as global meaningful actors (Leon et al, 2001; Ó Siochrú & Girard, 2003), while mastering their capacity of intervention in global politics.

Assuming that new 'forms' of politics, and new modes of political communication (collective and trans-national) are emerging in the global context (Arts, 2003), we believe that all the aforementioned aspects should be taken into consideration in order to develop a better understanding of the possible impact of non-state/public-interest actors on the world scene.

An Interesting Case Study: The World Summit on the Information Society

The UN World Summit on the Information Society[2] offers a meaningful opportunity to observe and analyse different aspects of the transformations concerning trans-national civil society organizations and their potential impact on global politics, particularly in the fields of communication governance and communication rights.

The first phase of the World Summit on the Information Society was held in Geneva, from 10 to 12 December 2003. This was the culminating event of a long preparatory process, composed of PrepComs and regional conferences, side events and related meetings[3]: eighteen months during which a number of officially recognized 'stakeholders' – governments, international organizations, business entities and civil society – have been involved in debates, negotiations, on-line and off-line exchanges and production of written documents.

The summit aimed at developing a common vision of the Information Society and drawing up a strategic plan of action for concerted developments toward realizing such vision. It also attempted to define an agenda covering objectives to be achieved and resources to be mobilized, within the framework of the UN Millennium Development Goals. The formal output of the process was a *Declaration of Principles* and a *Plan of Action*: texts that have been negotiated during the preparatory process by governmental delegations (with the written and oral contribution of other stakeholders) and adopted on December 12, 2003. The second phase closes in Tunis, in November 2005.

Civil society presence and participation in WSIS has been one of the main novelties in the first phase, which makes it a meaningful case study towards a better understanding of non-governmental actors' relevance in global politics, recalling that WSIS was *'the first time (in which) civil society has come together in such diversity and is such numbers from all over, to work together on information and communication issues'*[4].

Throughout the WSIS process a number of catchwords emerged, among which connectivity, development and digital divide. We suggested elsewhere that 'convergence' should be added as an underlying conceptual nexus:

> *... convergence not only in technology, but also in policy-making, actors' orientation and in discourse. WSIS has offered a window of opportunity to collectively refine the political agenda for communication, for policy-makers as well as for other 'stakeholders' and scholars. A content-oriented agenda, but also a process-aware agenda, which makes it relevant to focus both on the content issues ... and on the procedural aspects: the overall political outcome that parallels the final written outputs* (Padovani, 2004: 187).

According to Seán Ó Siochrú, WSIS has been also the convergence of two strands of debate in the history of communication politics:

> *One we term the 'information society' debate, taking in the role of information, the Internet and the 'digital divide'. It can be traced to the 1970s but the current manifestation found its defining moment in the mid-1990s. The other we term the 'communication debate', encompassing broader issues of knowledge ownership and use, media diversity and communication. It goes back as far, but its defining moment came in the early 1980s with the MacBride Report of UNESCO* (Ó Siochrú, 2004: 203).[5]

Given the focus of the present article, we suggest that WSIS can also be considered as the occasion for the convergence of (at least) two different 'realities' of global civil society.[6]

Converging Realities of Civil Society at WSIS

In reviewing efforts made to re-theorize democracy in the context of globalization, Catherine Eschle (2000) identifies three theoretical approaches to global civil society that have developed in the last decade: that of the cosmopolitan liberals, of which the work by David Held is a well-known example but which also relates to the work of the Commission on Global Governance,[7] which insisted on the role played by civil society and particularly formally recognized non-governmental organizations (NGOs) within it. The second conception is that of global Marxists, such as Robert Cox (1993) and world-system theorists (Wallerstein, 1990). The third one is the approach developed by post-Marxists (Falk, 1987; Lipschutz, 1992) who, inspired by 'new social movement' theory, *'argue for a revitalization of civil society as the core of a new radical democratization project'*. In this last version global civil society appears constituted by diverse trans-national social movement activities, and a crucial element is that it can be considered both *'as a terrain of democratization, with movements seeking to democratize relations within it, and as a source of democratization, with movements located within it seeking to constrain and transform the power of the state system and the global economy'* (Eschle, 2000).[8]

We argue that both the first and third 'conceptualizations' of civil society recalled by Catherine Eschle were at work in WSIS. We call the first approach 'institutional',[9] referring to the tradition of relations between the United Nations system and non-state actors, mainly non-governmental organizations (NGOs). A tradition, which – stemming from a state-centric perspective of international political processes – has certainly gained strength in the last decade, on the occasion of UN conferences (Pianta, 2001; Klein, 2004). Yet, such approach is now being challenged by the other, which we refer to as a 'globalization from below' approach, thus underlying its prevailing spontaneous character, its networking mode of operation and its 'bottom-up' implications.

As far as the institutional approach, the UN has a long history of relation with non-state actors (Ó Siochrú, 2002) that dates back to article 71 of its funding Charter. Rules to regulate interaction with civil society actors were afterwards adopted by ECOSOC in 1950 (Res. 288B) and 1968 (Res. 1296), and redefined in 1996 (Res. 31). The mid-1990s was the time of the growing visibility of NGOs and their growing presence at UN conferences that started with Rio in 1992, and proceeded to Vienna (1993), Cairo (1994) and Beijing (1995). During the Rio conference a first attempt to define the boundaries of the complex reality of 'civil society' was carried on through the identification of major groups, including gender, indigenous people, professionals, NGOs; while the Commission on Global Governance was also considering such developments in its investigation and proposals.

In 1998 UN Res. 53/170, speaking about civil society organizations, stated that they could *'no longer be seen only as disseminators of information, but as shapers of policy and indispensable bridges between the general public and intergovernmental processes ...'* This path, together with a growing awareness of the need for a democratization of the UN system through a more open and participatory functioning, led to the recent work of the above-mentioned High Level Panel on UN–civil society relations, and its final output: *'We, the People: Civil Society, the United Nations and Global Governance'* (June 2004).

This institutional approach to civil society landed at WSIS through Resolution 56/183, which encouraged *'intergovernmental organization, non-governmental organizations, civil society and the private sector to contribute to, and actively participate in, the intergovernmental preparatory process of the Summit and the Summit itself'*. Throughout the WSIS process the formula adopted was 'NGOs and civil society', thus differentiating between the two and recognizing that civil society is something different (and, as the process demonstrated, less defined) than NGOs. Nevertheless we suggest that the underline conception of civil society actors, characterizing governmental delegations and IGOs, remained an NGOs-based one, close to what expressed in ECOSOC resolution 31, where NGOs are described as not-for-profit entities whose *'aims and purposes shall be in conformity with the spirit, purposes and principles of the UN Charter'* operating at national, regional and international level.

But while the idea of a 'tripartite' mode of interaction was gaining momentum inside the UN,[10] from the early 1990s with the global emergence of the Zapatista movement, and subsequently even more visibly with the Seattle WTO meeting in 1999, something started to change, not just in the streets of cities like Genoa or

Cancun, but in the media, in common discourses, in trans-national organizations' everyday life, on the Internet, due to the fact that globalization processes were being de-constructed from below and the very legitimacy of international institutions openly put in discussion.[11]

Building on historical precedents, such as the anti-slavery movement at the end of the 19th century, on the expansion of connections among national social movements in the 1970s (Della Porta & Kriesi, 1998) and on relatively more recent forms of protest, like demonstrations against the World Bank in Germany just before the fall of the Berlin Wall (Keck & Sikkink, 1998), together with a growing awareness and competence in the use of long-distance communication devices, meaningful developments have taken place in civil society trans-national modes of organization. This has led to experiences such as the World and Regional Social Forum meetings, which can be conceived as networks of networks (Della Porta & Mosca, 2004); to local and national gatherings aiming at building alternative and independent visions for globalization processes; to occasions for trans-national connections to be created, experienced, strengthened and communicated.

Interestingly, not only 'traditionally central' issues related to globalization are debated in such spaces. Also an apparently only-for-expert topic, such as communication rights, has slowly gained its place in the agenda. From the 2002 (second) edition of the WSF, communication started being dealt with no longer just in instrumental terms – how should communication and information technologies be used as tools for internal organization and external outreach – but also in substantial terms.[12] The need for a democratization of communication, the implications of convergence in the global ownership of communication and topics such as how should ICT be governed in order to promote a more democratic international system, are all issues that contributed to relate communication and information to the wider 'globalization from below' discourse.

A growing attention posed on communication and information issues, together with the opportunity offered by the upcoming World Summit, allowed a number of individuals and networks, which had been active for years in the promotion of an open and democratic use of 'old' and 'new' communication technologies, to find new motivation and energies to come together and become active inside the WSIS preparatory process.

Thus the 'globalization from below' vision and practice of global civil society also landed at WSIS: a reality which, according to some, has historical and conceptual roots in the NWICO debates of the 1970s (Traber & Nordenstreng, 1992; Nordenstreng, 1999; Ó Siochrú, 2004) and had developed its own networks and strengthened its international visibility in the 1980s and 1990s through loose initiatives such as the MacBride Roundtables, the proposal for a People's Communication Charter or the Platform for Democratic Communication; but also through projects for development cooperation in the field of communication for social change (WACC) as well as through more formalized structures such as the World Association of Community Radio (AMARC) and the Association for Progressive Communication (APC). Overall, the network recently re-asserted its identity through the launch of a Communication Rights in the Information Society campaign.[13]

Furthermore, at WSIS this 'tradition' of mobilization on communication and cultural issues met with more recent experiences related to the use of ICTs for an 'Internet citoyen and solidaire', with the open source movement, with the fragmented yet very active reality of the digital rights movement, with interesting examples of 'globalization from below' such as the Global Community Partnership.

We can say that Geneva has favoured the gathering of different experiences of social mobilization on communication and information issues, from the most 'ancient' to the most recent, stemmed from the evolution of ICT use among citizens and communities. At the same time WSIS has contributed to a dialogue among associational structures of civil society that are more institutionalized and 'expert' of global processes – such as the CONGO or a number of NGOs that were induced into the process through UNESCO and its own networks[14] – and more spontaneous forms of mobilizations that are the expression of the articulated galaxy of the global movement for social justice. An unprecedented plurality of actors and discourses, of visions and modes of interaction with institutional actors which deserves further investigation. Some 200 accredited non-governmental entities took part in PrepCom1 (September 2002), some 1500 civil society entities have registered in the official civil society website (www.geneve2003.org) and over 3300 participants from non-governmental organizations and civil society attended the first phase of the summit.[15]

The outcome of such encounters were not foreseeable at the start: the different experiences and political cultures represented by such a diversified reality could have produced fragmentation and conflict, in developing discourses, elaborating documents and defining political practices; as well as it could have fostered partial convergences or an homologation on the positions expressed by stronger actors. Our analysis, and personal observation, suggests a different result: a meaningful convergence, through a process of collective elaboration, in the respect of the plurality of voices and positions.

Visions and Convergences: Perspectives on Governance

Theoretical conceptions of global civil society obviously draw on the observation of practices that have developed over time. Hence it is important to note, as we have done above, how civil society organizations' praxis in the trans-national environment today presents a plurality of manifestations of formal and informal character, institutionalized relations as well as spontaneous self-organization, habits of dialogue with formal institutions together with strong expressions of contentious politics. Such plurality is a crucial element, since different realities of civil society, their nature and the role they perform, may give way to different perspectives, and possibly praxis, of global multi-actor governance.

At the same time, it is also important to stress the role of discourses: the conceptualization, self-perception and representations that both civil society and other subjects are developing, will contribute to the re-definition of state and non-state actors' role in the international arena.

In our empirical analysis of WSIS we are mainly focusing on the construction of narratives, referring to other authors for an in-depth investigation of internal relations within the civil society sector as well as for the analysis of their interaction with governmental actors (Raboy, 2004; Ó Siochrú, 2004; Kleinwächter, 2004; Cammaerts & Carpentier, 2004). We are interested in the outcome of the convergence among different realities of civil society in terms of 'visions'; and therefore we look at two different semantic spaces: that of internal dynamics within the civil society sector and that of interaction with the official process.

Moreover, since Geneva has been the occasion in which different actors, while participating in a process concerning information and communication issues, have also made explicit their understanding of governance processes in a globalized world, we shall focus precisely on the different conceptualizations of political processes that emerge from the documents, within the wider discourse on information and society.

We can summarize the plan of our investigation as follows:

CS convergences @ WSIS	Practice[16]	Visions of governance
within civil society	who what how	Lexical-textual analysis of civil society documents[17] throughout the preparatory process. Focus on 'governance' (the who, what and how of global governance within the civil society sector discourse) to evaluate the degree and evolution of internal consistency in a processes that involved a plurality of voices.
CS relation to official process	who what how	Lexical-textual analysis of official WSIS documents together with those elaborated by the civil society sector in two stages of the process: PrepCom 2 and the final summit. Focus on 'governance': the who, what and how of global governance comparing official and civil society visions.

We broadly refer to 'governance' as a 'process of interactions among different actors at different levels' for the definition of rules and lines of conduct (Padovani & Nesti, 2002). Governance has been a constant focus in our investigation of WSIS as a political process and we did not just focus on the explicit use of the term or the specificity of ICT and Internet governance: we concentrate on the 'inner vision of steering processes' that can be derived from written texts.

Applying lexical-textual analysis to investigate governance,[18] we have selected and tagged a number of 'complex textual units' (CTU) referring to 'actors' (actor/s, party/ies, stakeholder/s, country/ies, nation/s, individual/s, people, cities, private sector, civil society organization, but also users, citizens, decision-makers, etc.) and 'levels' of political action (national, regional, global, national and international, etc.). Moreover we identified units referring to 'modes of inter-action' and evaluation (cooperation, benchmarks, consultation/ing, outcome/s, commitment/s, regulation/tory, etc.) and units referring to the 'quality' of governance (democracy, democratic, participation/patory, empower/powering, partnership, openness, transparency, representative/ness, competitive/ness and the like). Actors and levels indicate something about the WHO in multi-actor/multi-level governance, while modes and quality tell us something about the WHAT and HOW of the process.

Civil Society Visions

Focusing on the dynamic internal to civil society, we were initially surprised to realize how little reference to the governance dimension could be found in early documents elaborated by the civil society sector, when analysed in relation to governmental texts, in spite of the efforts made by some civil society actors to contribute in the definition of a model for the multi-stakeholder process from the very beginning. After investigating documents from PrepCom2, in July 2003, we wrote: *'It seems that civil society actors, being mostly concerned with the affirmation of principles and values and with the possibility of widening the WSIS agenda, are not sufficiently focused on how "information_and_communication_societies" should be steered and regulated: very little reference to actors' role is made in comparison to other documents. No specific interest for governance emerges from civil society contributions. We can, probably, expect more indications about regulations and framework to come from civil society actors, in subsequent stages of the WSIS process.'* (Padovani & Tuzzi, 2003)

It was therefore interesting to discover, at the end of the process, that the governance dimension had not just been developed by civil society in the alternative Declaration adopted by the Civil Society Plenary on December 8th: the vision that emerged from that document was also very precise in its determination – about the who, what and how of governance – and sensibly more articulated and comprehensive than the one expressed by the official documents (Padovani & Tuzzi, 2004).

We therefore decided to reconstruct the learning process that led to such changes. We selected all the 12 documents elaborated by the Coordinating /Content and Theme Group of the civil society sector and clustered them according to the seven phases of WSIS.[19]

From the entire corpus vocabulary we selected 230 CTUs relating to governance,[20] and analyses were made referring to 165 CTUs with frequency above three. Within civil society documents, of all these units only 35 were specific to some phases (relatively more important in comparison to other, yet utilized also in other phases) and 58 were exclusive units (utilized only in specific phases). All other governance units were quite evenly utilized throughout the process in civil society documents. Amongst these: civil society (recurrence: 146), public (80), policies (59), people (57), national (49), citizens (40), implementation (40), governance (39), framework (38), private sector (37), all-stakeholders (32), transparent/transparently (respectively 23 and 23). This means that a civil society vision of governance has accompanied the entire process, presenting different elements – actors, modes and quality – of governance processes. In spite of the plurality of actors, and converging realities of civil society, the governance discourse developed by civil society shows a significant consistency over time, particularly from the Informal meeting (November 2002) onwards. After the first preparatory meeting in Geneva,[21] therefore possibly influenced by the direct experience of the process, a coherent 'vision' of governance started to emerge from the diversified realities that gathered around the Plenary and Content and Theme Group.[22] The basics of such vision can be found in the Informal meeting document. Here the 'actor' element is plural (civil society organizations, all stakeholders); the 'how' opens up to issues such as responsibility, partnership, decentralization and empowerment, anticipating themes that became central afterwards; the 'what' element also shows a pragmatic approach: mention is made of regulation, best practices, outcomes, enforcement, implementation. Internet governance also emerges as a theme in this early stage.

No exclusive CTUs are found in this phase, which set the common ground for the development of a more articulated discourse. In fact when we look at specific and exclusive units in the corpus we do find variations in the different phases: these can be related to specific events and stages of negotiations inside the official WSIS, to which civil society documents were reacting, but can also be considered as part of a broader learning process of consensus building through which civil society developed its own perspective. This explanation seem to be sustained by the fact that the emerging 'vision' from the final document is the most articulated and balanced one; it is also the document in which we find the highest number of exclusive governance units (27 out of 58).

Figure 1 shows the positioning of civil society documents, clustered according to WSIS phases and their use of governance CTUs.

Figure 1: Built on governance CTUs. Positioning of civil society documents, clustered according to WSIS phases; visualization of governance CTUs that most contributed to the determination of the axes

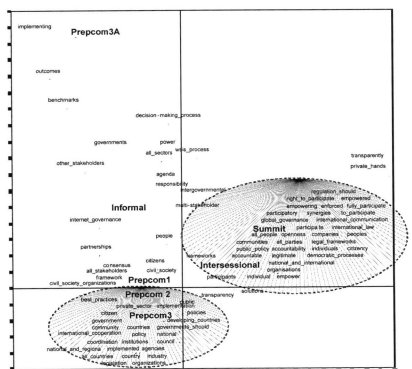

Documents from PrepCom2 are visibly situated in the middle of the graph, which means there are few units that are either specific or exclusive to that document.[23] Nevertheless PrepCom2 is richer in governance language than former phases. As far as the 'who' of governance, it shows an unprecedented attention for regional situations and developing countries (least developed countries, developed and developing, regional level, north–south, south–south) but it also introduces the local level of authority (local authority/governments). This can be interpreted as a 'localization' and a 'specification' of relevant actors, both in terms of institution and in terms of their intervention.

As far as the 'how' of governance, democracy and participation, as well as empowerment and decentralization, have gained more importance (through the use of different expressions: participate, fully participate, right to participate). The few exclusive CTUs tend to strengthen the value dimension (democratically, unaccountable); but the 'what' is also there, and solutions, legal and regulatory frameworks, commitments are mentioned, together with the first reference to good governance.

PrepCom3 is strongly focused on actors and levels, in their complexity: governments and local authorities, industry and regional and international level are specific to this phase. It is also interesting to see what actors' units are exclusive of these documents: private sector and civil society, public and private sector, multilateral/international organizations, all indicate a clear awareness of the multi-actor and multi-level nature of governance processes; a vision in which non-state actors are always mentioned together, private entities and public interest groups. This finding goes together with a strong self-reference to civil society operating inside the WSIS process (all caucuses and working groups – African, human rights, Latin America, gender, youth, community media – and WSIS-civil society are continuously referred to); which can be explained by the fact that some of these documents are working papers, in which self-reference to the 'author' is recurrent.

No specific unit concerning the 'how' of governance is found in this phase, while exclusive are only 'effectiveness' and 'legality'. This does not mean this element is absent; reference to transparency, responsibility and accountability is there, maybe not as relevant as elsewhere. The real novelty from PrepCom3 documents is the explicit mention of the term 'power' which becomes central towards the final stages of the process.

The 'what' element is also there and pragmatically developed: policy, solutions and regulation go together with best practices, outcome and governance which is declined in different ways: ICT governance, Internet governance and, again, good governance.

In figure 1, the summit declaration is positioned in a space opposite to PrepCom3. This can be explained, again, by looking at specific and exclusive CTUs. As far as actors are concerned it is interesting to note, together with a very inclusive approach (all actors, all citizens, all people), a strong focus on communities and peoples (always plural) on one side and on the international community on the other. This suggests a parallel, implicitly made by civil society, between traditional actors in world politics (the international community composed of states) and non-traditional actors (communities and peoples), which aspire at being recognized. States, governments and countries are still mentioned, but the relative importance of the international *versus* grass-roots communities seems to stress the contraposition between old conduct of world politics and the new governance, which is needed for the 21st century.[24]

It is also worth noting that the final declaration is the only text in which civil society is always referred to as 'global': the 'author' is no longer considering its action as confined within the WSIS process. Global civil society is a strong statement that underlines actors-within-WSIS' sense of belonging to a wider global constituency. These two elements show that the self-referring tendency of former phases has given the floor to a more comprehensive, and cosmopolitan, understanding of civil society. And global has also become the very concept of 'governance' (global governance, ICT global governance).

Few specific units in this document refer to the 'what' of governance, if not for a strong reference to international law and regulation, suggesting that decision-making and public policies should be developed within legal regulatory frameworks. Redistribution, reinforcement and reform are exclusive units to this text.

As far as the 'how' or 'quality' of governance is concerned, two aspects should be mentioned. The participatory dimension, which has accompanied the entire process, reaches its highest point in this document: together with participation, participatory, full participation and the like, we find a stronger 'right to participate'. This goes together with a second interesting element: not just empower, empowerment and empowering, but the very concept of 'power' (which had appeared once in PrepCom3) is utilized three times in the text and exclusive mention is made of powerful and unequal power.

We suggest a connection between these two elements: having been involved in the WSIS process for 18 months, civil society actors developed a clearer (more realistic?) understanding of global civil society involvement in world politics. An understanding that is aware of the difference between being able/invited to participate as a stakeholder and having the possibility to exert some 'equal power' at the global level (Cammaerts, 2004; see alsoCammaerts & Carpentier in this volume). This would support our belief that the 'multi-stakeholder approach' is not yet a model and needs to be defined, not only in theory but in practice, taking into consideration the nature and level of power that different stakeholders can exercise.

As far as civil society realities convergence at WSIS, we believe that the coherent evolution of a civil society 'vision' of governance throughout the process, being the result of a collective exercise participated by a number of different civil society actors, indicates a positive outcome in their convergence: the result of negotiation processes, effectiveness of mechanisms for consultation and consensus-building, capacity to develop a common and agreed upon language. The *Civil Society Declaration* confirms the strong focus on values (transparency, accountability, responsiveness) and norms (legal framework, regulatory aspects), which prove to be a basic common ground for civil society actors.

Civil Society and the Official Process: Comparing Narratives
If we now focus on the relation between civil society convergence and the official process, we can underline the difference between governance visions expressed by

civil society documents and those of the official texts. Building on former analyses (Padovani & Tuzzi, 2003; 2004), we here offer an overview through graphs in which documents are positioned in the WSIS semantic space at PrepCom2 (February 2003) and at the summit (December 2003).[25]   Again, governance CTUs are visualized.

Figure 2: Positioning documents in the WSIS semantic space at PrepCom2 (selection of texts from different actors: official, regional/governmental, private sector, civil society). Visualization of governance CTUs that most contribute to axes definition. From Padovani &Tuzzi (2003)

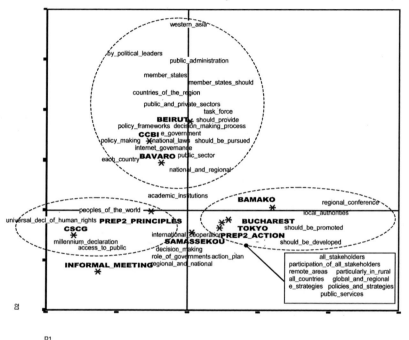

In the WSIS space of discourse at PrepCom2, three semantic areas[26] indicated different visions about who should contribute (and how) to the definition of a regulative framework for the information society. The official *Declaration of Principles* is being positioned close to the contribution elaborated by the civil society. This can be explained considering the common priority given by the two documents to the value dimension in that stage of the process. In comparison with the language of all other texts, none of the two seemed to express any specific understanding neither of governance nor of the role of actors within it. But the official process actually developed around the right and upper semantic areas, since a compilation of Reports from the regional conferences and a 'non-paper' elaborated by the President of the preparatory process were the actual basis for subsequent negotiation. Thus the official process 'vision' should be drawn by documents such as the reports from the Bucharest and Tokyo regional conferences and the *Plan of Action* (centre to right area), characterized by sequences such as 'participation of all stakeholders' and 'stakeholders should', together with reference to the different levels of action. An idea of multi-actor and multi-level governance emerged from those documents, while no reference to the 'how' and 'what' of governance was central to those texts.

Interestingly, the CCBI (private sector) document and Beirut and Bavaro regional reports (upper area) suggested a quite different idea of governance: no reference to the plurality of stakeholders, and a specific use of units such as 'by political leaders', 'member states', 'states should' and 'all countries' to indicate a strong focus on institutional regulation, in an environment where state actors still have a crucial role to play.

Within the official process at PrepCom2, perspectives on governance were diversified amongst governmental actors themselves and a clear distinction already appeared between the governmental and civil society documents.

After the final stage of the summit, we conducted an analysis of all final documents presented in Geneva on at the closing session of WSIS, December 12th: the *Declaration of Principles* and *Plan of Action* together with seven other documents elaborated by 'civil society' actors. Again visions of governance were investigated. Figure 3 is a visualization of governance CTUs in the general semantic space of the summit.

Figure 3: Positioning final documents in the WSIS semantic space and visualization of governance CTUs that most contributed to axes definition. From Padovani & Tuzzi (2004a)

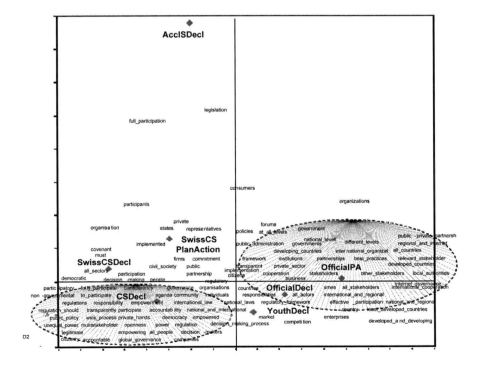

From the analysis it becomes clear that the plurality of visions about the information society is also a plurality in governance understanding. All documents refer to at least some of the elements we have selected to identify visions of governance: actors, levels, modes and quality. Here we only recall some findings concerning the official documents and the alternative declaration adopted by the Civil Society Plenary.[27]

The official documents are characterized by a very specific definition of actors and level: countries are developing, developed, least developed; action takes place at the national, national and international, national and regional levels. Cooperation is regional and international. The regional dimension appears strongly only in official documents, a strong way of conceiving multi-level governance that is not shared by other actors. Stakeholders are important (they are relevant stakeholders, multi stakeholders or other stakeholders) as subjects to build partnership with. But they are hardly identified: no cities and local authorities, little civil society, little communities and peoples; a stronger focus being on private sector, business, firms, SMEs.

Governance is either 'good governance' or it relates to Internet, thus gaining specific meanings. Very little is said about the modes and quality, the 'how' and 'what': a part from a strong focus on competition and competitiveness, we find little reference to democracy or empowerment. Participation is there but it is not central, nor does the idea of regulation or decision-making appear to be central; a generic 'regulatory framework' formula prevails.

On the other side, what characterizes the Civil Society Plenary document is, as indicated above, a very inclusive approach (all actors, all citizens, all parties, all peoples) combined with the emphasis on participation, which is expected to be 'full' more than 'effective' and, as noted above, a basic right. A strong vision of the quality of governance emerges, with reference to actors (decision-makers and international community but also citizens and people, civil society organizations, communities and private sectors); tools that should be developed (agenda, sanctions, regulation, covenants, enforcement, international laws); and the quality of such governance (democratic, legitimate, empowering, accountable, transparent).

We can say that what was already a distance in language and in the conceptualization of governance, between the official process and civil society at PrepCom2, has possibly widened by the end of the process. Convergence among realities of civil society might have contributed to define a common vision of governance within the civil society sector, but it did not contribute to bridge the distance between the top-down/governmental and bottom-up approaches that met at WSIS.

Conclusion

The World Summit on the Information Society has certainly been a complex event; and actors' participation has been equally complex. As far as civil society is concerned, there has certainly been a convergence both in practice and visions, though never a complete one, nor was it desirable. The plurality of actors and positions and the plurality of final documents attest once again the complexity of a reality, which is too easily labelled as 'trans-national civil society' and simplified, sometimes favouring criticism.[28]

Nevertheless, the articulated organization and self-structuring that civil society actors have developed during the process, were necessary to play a role, and promoted a process of convergence especially among those two 'realities' of civil society that have amongst their repertoires of action precisely the organization in/of parallel summits: NGOs and activists from social movements (Kaldor, 2003: 80–81). This link could be understood as the development of a 'trans-national civic network' or a 'trans-national advocacy coalition'; which are defined as networks that connect NGOs, social movements and grass roots organizations (Keck-Sikkink, 1998). Some of these subjects being closer to institutional settings, others more activists, acting together in a form of cooperation where *'the latter tend to be more innovative and agenda-setting, while the former can professionalize and institutionalize campaigns.'* (Kaldor, 2003: 95)

We believe that two novelties should be underlined in the case of the WSIS. The first one is that civil society presence at WSIS was not in the form of a 'parallel summit'. As we mentioned, NGOs and civil society where invited to participate and they did so, in the very same physical space as the official summit, making the effort to continuously relate to the official intergovernmental process, while at the same time developing positions and organizing their own channels for exchange and cooperation. If such involvement was satisfactory is a matter for further discussion; nevertheless WSIS has set a precedent in the history of global politics, while showing the difficulties, potentialities and shortcomings of a new approach to global governance.

The second relevant aspect concerns the content dimension. The kind of convergence that took place at WSIS cannot be defined as an 'advocacy coalition', since coalitions normally concentrate on single issues. The WSIS process has in fact witnessed the dialogue between activist, hacktivists, grass-roots groups, exponents of epistemic communities, individuals and NGOs, the former being more creative and agenda-setting-oriented and the latter extremely helpful in mediating the formal presence of civil society in the process. All those actors were concerned with the most differentiated issues, from media concentration to open source, from ICT for development to people with disabilities, from technological waste to human rights, from gender issues to indigenous peoples, from global justice to the empowerment of communities through knowledge. A plurality of issues was brought on the agenda since the challenge was to build 'visions of (information and communication) societies'.[29]

We therefore believe that not a just a trans-national coalition but a global dynamic of social movement was in action at and around WSIS, an hypothesis that seems to be sustained by the continuity of exchange that have followed the Geneva event through Tunis and is ongoing, at different levels in different forms, mainly but not exclusively mediated by long-distance communication tools.

In terms of visions of governance, and the role of civil society as an actor within such vision, what emerge from our analysis is that in the official/governmental perspective, the governance landscape mainly concentrates on the map of actors and levels: actors (always considered in a 'macro' dimension) are specified, countries are declined, levels of action are articulated; while very little mention is made of the very nature of governance, which would define the role and position of the different actors. They appear as juxtaposed but not interacting. We suggest that the outcome, in terms of governing style, would be an 'aggregative mode' of governance,[30] in which actors play a role on the same scene though not necessarily building dialogues. This aggregative mode can be thought of as a negotiation in which actors, following different logics of exchange and mediation, participate in policy-making with very different power resources. At the global level, this reflects the long legacy of diplomacy styles. Within WSIS this legacy has strongly informed the official process, thus defining the 'official multi-stakeholder approach' as an aggregate of actors exerting different power and playing different roles.

In contrast, the bottom-up perspective that characterizes the declaration elaborated by civil society shows little interest for the mapping exercise. What matters at the grass-roots level, to activists as well as to NGOs, is the 'how' of governance: responsiveness and accountability of institutions and empowering participation of actors. This would suggest a more 'integrative approach' to governing modes, based on the negotiation of interests (and words), through dialogues that allow each actor to redefine its priorities as well as its identity. This 'concertation exercise' implies that specific interests are re-elaborated in order to reach a common consensus. From our findings and observations we can say that it has been precisely an approach of this kind that was adopted within the civil society constituency; and that has proven to be possible within a trans-national political process, though only within a specific sector and not in the interaction with actors of a different nature (governments and IGOs).

Even when considering the number of contributions developed by different civil society groups (figure 3) – which attests of the persistence of a plurality of civil society positions and convergences – what is to be noticed is the fact that all civil society documents are positioned on the same side of the graph, above or below the left end of the horizontal axis. That can be viewed as the semantic area of the 'how' of governance, where units that indicate the 'quality' of processes define the basics for effective and meaningful participation of different stakeholders.

A final consideration should be made in terms of the 'impact' of civil society participation. We suggest that not only the output of the summit – the final documents – should be evaluated, but particularly the outcome, which is the overall political process as a learning space; an evaluation to which we hope to bring a contribution through our reflections. Given our focus on civil society as a global actor, in talking about 'outcome' we refer to the different results (some of which unexpected) of civil society involvement. Among these: a contribution in broadening the agenda, a fruitful convergence of different civil society actors, and a continuity of interactions beyond the WSIS process.

To conclude, WSIS has shown the articulation of civil society realities and the multiplicity of networks and connections that can develop from interaction in a common space, which is no longer just physically defined but complemented through long-distance connections. In spite of this complexity, it seems that amongst civil society organizations a strong and shared awareness has emerged: only through the development of a qualitative dimension of governance can a discourse on non-state actors participation in global politics be elaborated beyond rhetoric and actualized. This can no longer be considered only as the result of the 'value orientation' that characterizes civil society. The pragmatic and substantial reference to international laws, legal and regulatory frameworks together with explicit reference to the power dimension suggests that there is more to civil society than 'just value' in its approach to global governance for the 21st century.

Appendix

Lexical-textual Analysis: Presenting the Method

Step 1. Evaluation of Dimensions
The corpus for content analysis is a collection of written texts organized according to a grouping criteria. A corpus is composed of words, which are only sequences of letters taken from the alphabet and isolated by means of separators (blanks and punctuation marks). A word-token (wto) is a particular occurrence of a word-type (wty) in a text. A token instantiates a type (so, for example, the single word-type 'the' has many tokens in any English text), but there are also many word-types that occur only once in a given corpus (hapax legomena). The entire corpus includes a total of N word-tokens (corpus dimension in terms of total occurrences). The frequency of occurrence of a word-type in a document is the number of corresponding word-tokens repeated in the corpus. The list of word-types with each frequency includes a total of V word-types (vocabulary dimension in terms of different word-types) and is the vocabulary of the corpus.

The Type–Token Ratio (obtained dividing the vocabulary dimension V by the corpus dimension N) and the hapax percentage (number of word-types that occur only one time in the whole corpus divided by the vocabulary dimension V) are measures of lexical richness and since a statistical approach makes sense only with large corpora, they are useful to decide if the corpus is large enough. If the Type–Token Ratio is less than 20% and the hapax percentage is less than 50% it is possible to state the consistence of a statistical approach (Bolasco, 1999). From the point of view of lexical richness we can see that short documents always show a rich language, which can be explained through the limited dimension.

Step 2. Lexicalization: From Simple Word-types to Complex Textual Units
In a first phase of analysis only simple word-types are chosen in order to evaluate the dimensions of the corpus. Then we identify a number of complex textual units (CTUs) in the vocabulary and recod the corpus accordingly (Bolasco, 1999; Tuzzi, 2003). Complex textual units are used: a) to increase the amount of information (complex textual units carry more information than simple word-types); b) to reduce the ambiguity of simple word-types (simple word-types are ambiguous because they are isolated from their context of usage).

In order to recode the corpus we need to identify in the documents: all multi-words; all sequences of words that gain or change meaning if considered as a block and, more generally, all sequences that make sense and are repeated several times in the corpus. This operation can be easily performed through the use of Taltac software (Bolasco et al., 2000). Using Taltac procedures we first obtain a list of sequences of word-types repeated in the corpus composed of several thousand of sequences. Since most of them are empty (*i.e.* 'and in a'), redundant (*i.e.* 'cultural and', 'cultural and linguistic', 'and linguistic diversity', 'linguistic diversity', etc.), or incomplete (*i.e.* 'persons with' or 'countries with economies in') we then select the most informative sequences according to the Morrone's statistical IS index (Bolasco et al., 2000), combining this with a manual control in order to obtain a new list of 'the best sequences'.

The final list of 'the best sequences' is used for the lexicalization of the corpus. This means that, for example, a repeated sequence such as 'countries with economies in transition' is re-written in the corpus as 'countries_with_economies_in_transition' and the sequence is, thus, recognized as a single complex textual unit. After the lexicalization procedure word-types, multi-words, poly-forms, idioms, and repeated sequences (all of which we define as 'complex textual units') appear together in the same new vocabulary.

Step 3. Selection of Complex Textual Units

Starting from the new vocabulary per CTUs and remembering that it contains either simple word-types (i.e. 'governance') or lexicalized repeated sequence (i.e. 'all stakeholders'), we select a sub-set of CTUs for the conduct of further analysis according to five criteria in a hierarchical order:

1) topic textual units. We tag some CTUs in the vocabulary in order to be able to control terms that were of interest for a specific topic useful for a thematic reading of documents (e.g. all CTUs concerning 'governance').

2) specific textual units. In order to recognize CTUs that are present noticeably more (or less) in a document than in the corpus as a whole, we use the traditional 'characteristic textual units' method (Bolasco, 1999; Tuzzi, 2003). This simple tool is based on the hypergeometric model and by means of a probability of over-usage it can detect which elements are used frequently inside a document. All CTUs that show a high probability of over-usage for a document (p less than 0,025) can be considered 'specific' for that document, which means peculiar to that document with reference to the others.

3) exclusive textual units. In order to assess the originality of a document with respect to others, we select all CTUs that are used in each document in an exclusive manner (they occur only in a document and never in the others).

4) repeated sequences. Starting from the list of CTUs that are neither 'topic', nor 'specific', nor 'exclusive', we focus on CTUs born from the lexicalization of repeated sequence, according to the same logic that led to our codification in complex textual units: multi-words, idioms and repeated sequences carry more information and less ambiguity than simple word-types.

5) frequency threshold. Since it is not possible to work with all the selected CTUs (still too many), it is necessary to set a consistent threshold and focus the analysis on CTUs with a frequency higher than this threshold.

However to conduct more qualitative and in-depth investigation all the CTUs contained in the vocabulary and also hapax should be considered.

Step 4. Correspondence Analysis

We build a two-ways contingency table with rows named with the selected CTUs and columns named with the grouping criteria (documents, authors, et.) where for each unit we can read in the cells how many times each author/document has used it.

In order to obtain a graphic representation of the contingency table we apply correspondence analysis (Bolasco, 1999). This statistical technique allows to represent the system of relations existing between authors/documents and selected CTUs on a Cartesian plan where each CTU and each author/document is positioned by means of coordinates. Such positioning is fundamental for the interpretation of the solution, because the most important CTUs for a author/document fall close to the author/document.

The entire system of relation contained in the two-ways contingency table can be drawn on a multidimensional graph in which each author/document and each CTU is a point in a hyper-space by means of coordinates. The comprehensive representation would be very complicated. It is therefore better to observe one axis at a time (one-dimensional point of view) or two axes a time (bi-dimensional point of view or dots on a Cartesian plane). Further difficulties derive from the number of CTUs we want to draw. They cannot be all represented on the Cartesian plane at the same time. For this reason we represent only those that are more important for the reading of the solution since they play a prominent role in determining the geometrical setting.

Application of the Method to Documents Elaborated by the Civil Society Sector within the WSIS Process

Step 1. Evaluation of Dimensions

Our corpus is composed of twelve documents written by CS actors and grouped according to the seven phases of the preparatory process of the World Summit: PrepCom1, Informal meeting, PrepCom2, Inter-sessional, PrepCom3, PrepCom3A and Geneva Summit. The corpus includes a total of $N= 53,949$ word-tokens and $V= 4,380$ word-types. The Type–Token Ratio $(T.T.R.=V/N= 8.12\%)$ and the hapax percentage (39.57%) allow us to state the consistence of a statistical approach.

The length of the seven clusters of documents is different: the longest is PrepCom3. The shortest is PrepCom1.

Step 2. Lexicalization: From Simple Word-types to Complex Textual Units

Using Taltac procedures we obtained first a list of sequences of word-types repeated at least two times in the corpus. Then we have selected the most informative sequences according to the Morrone's IS index together with a manual control and obtained a new list of over 1,000 'best sequences' useful for the lexicalization of the corpus. The dimension of the corpus after this recoding procedure is $N= 47,464$ occurrences and the dimension of the new vocabulary is $V= 5,023$ CTUs.

Step 3. Selection of Complex Textual Units

We have selected from the CTUs vocabulary a sub-set according to the five criteria (topic textual units, specific textual units, exclusive textual units, repeated sequences and frequency threshold). For further analysis applications we have decided to use only forms with frequency higher than 2 (fixed freq greater than=3), meaning a sub-set of 1,570 CTUs.

Step 4. Correspondence Analysis

We have built a two-ways contingency table with 1,570 rows named with the CTUs and 7 columns named with the WSIS phases. From this contingency table, correspondence analysis obtains six axes and in the graphs we have visualized the first two. Furthermore we have decided to represent separately graphs concerning the governance theme.

All the graphs that are shown represent a percentage of explained inertia by the first two axes higher than 50%.

Notes

1   'We, the People: Civil Society, the United Nations and Global Governance' (UN A/58/817, June 2004).

2   Relevant links: www.itu.int/wsis; www.wsis-online.org; www.geneve2003.org; www.worldsummit03.de.

3   Complete reconstructions of the process with insights and related documentation can be found on the official website – www.itu.int/wsis – but also in the Report elaborated by Raboy & Landry: La communication au coeur de la gouvernance globale available at www.lrpc.umontreal.ca/smsirapport.pdf.

4   Seán Ó Siochrú's speech at the WSIS Plenary, December 11th.

5   In order to outline the historical legacy between former international debates on communication issues and recent developments, we have applied lexical-content analysis to the final documents from Geneva (the Official Declaration and the alternative Declaration written by the civil society group) and to the final recommendations expressed in the MacBride Report (1980), with the aim of tracing changes and continuity in language and content (Padovani & Tuzzi, 2005).

6   Roberto Savio speaks about 'generations' of global civil society, referring to developments from the early 1990s, and the presence of NGOs in UN World Summits (which can be considered as one of the outcomes of the trans-nationalization of social movements dynamics from the 1970s which focused on issues such the environment, peace and human rights or gender issues); to the Seattle mobilization and follow-ups, from Stockholm to Genoa; to the 'new' World Social Forum environment which is understood as a space of complex dynamics, characterized by a higher degree of autonomy from institutional settings (intervention at the Euricom Colloquium Information Society: Visions and Governance, Padova – Venice May 2003). Differently, Mary Kaldor (2003) writes about 'versions' of global civil society, underlying how mobilization phenomena have bee referred to over time, since the 18th century, by different subjects (thinkers, institutions, actors from the civil society themselves). In a yet different way, Catherine

Eschle (2000) presents a series of 'conceptualizations' about civil society, looking at democratization theories from the 1990s. We prefer to adopt the expression 'realities' of civil society, in order to stress that, in spite of chronological developments, we now witness a co-presence of different expressions of global civil society organizations, acting today on the world scene. Moreover we are looking at concrete modes of political participation.

7   *Our Global Neighbourhood*, Final Report (1995), Oxford University Press.

8   Mary Kaldor further differentiates between a 'post-modern version' and an 'activist version' of civil society, underlying the plurality of global networks of contestation that characterizes the first and the focus on the emergence of a global public sphere which is peculiar in the conceptualization of the second.

9   Both Eschle and Kaldor refer to this as 'the liberal vision' where 'civil society consists of associational life – a non-profit, voluntary third sector – that not only restrains state power but also actually provides a substitute for many of the functions performed by the state' (Kaldor, 2003: 9). We adopt the label 'institutional' in order to stress the conception of civil society as composed by formal, identifiable organizations; the top-down character of such a vision, elaborated by institutional actors searching for 'representative' and identifiable interlocutors, and its consequences in terms of an 'aggregative' model of governance, which will be discussed in the conclusion of this article.

10 Particularly innovative in this sense have been certain UN agencies and programmes, such as the International Labour Organization, UN Habitat, UNAIDS, where actions were taken to foster and formalize consultative mechanisms involving governments, private sector entities and civil society as the three 'parties'.

11 Defining the time frame for global social transformation is clearly an arbitrary exercise. As far as trans-national connections in the post cold war era are concerned, we tend to agree with Manuel Castells (2000) in considering the role played by the Zapatista insurgence in the early 1990s and their innovative use of Internet and ICTs, as turning points (Padovani, 2001).

12 This has been investigated by Stefania Milan in a thesis on *Civil Society Media at the WSF* and then compared with initiatives within the WSIS in Hinz & Milan, in a paper presented at the IAMCR Conference in Porto Alegre, July 2004.

13 Web references for above-mentioned initiatives and associations: MacBride RoundTables www2.hawaii.edu/rvincent/macbride.htm; People's communication charter www.pccharter.net/chartere.html; WACC www.wacc.org.uk; AMARC www.amarc.org; APC www.apc.org; CRIS campaign www.crisinfo.org.

14 We recall that UNESCO has been the most active international organization in consulting with civil society, as demonstrated by the meeting organized in Paris, in April 2002, before the formal start of the WSIS process, which contributed to defining UNESCO positions within the process; as well as by the on-line consultation conducted in December 2002, see http://www.unesco.org/wsisdirectory.

15 For a thorough analysis of civil society participation in WSIS, see Cammaerts & Carpentier: *The Unbearable Lightness of Full Participation in a Global Context: WSIS and Civil Society participation*, in this volume.

16 We refer to the several and detailed reports on the role performed and the structures developed by the civil society sector at WSIS (Raboy & Landry, 2004; Ó Siochrú, 2004a). We also refer to the website set up by the civil society sector to organize its structure and communication channels: www-wsis-cs.org. Here we only recall, to set the context for a better understanding of our investigation, that the ultimate authority for the civil society sector at WSIS was the Plenary Assembly, which would take collective decisions, also concerning the written work elaborated and coordinated by the Content and Theme group (building upon a number of working groups, set up according to thematic focus or geographical representation).

17 Most of our analyses refer to documents that have been elaborated by the Coordinating Group of Civil Society (CGCS, later Content and Theme Group, CT), linked to the Plenary. The reasons for this choice are the following: that was the group which allowed the widest collective cooperation within civil society and it would have been impossible to track all documents presented by all civil society accredited entities. Furthermore, given our interest for convergences inside the WSIS, we should recall that the CT group has been recognized, from the beginning, the ultimate competence in terms of content development in the name of civil society. It worked through a core group of people (volunteers) who coordinated the efforts made by several caucuses and working groups, which were the actual spaces where thematic debates took place amongst subjects coming from the most different backgrounds (NGOs, professional, researchers, campaign exponents...). Consultations were done both on-line and off-line and consensus was reached on specific formulations referring to single aspects; such formulations were then channelled through the CT group and contributed to the documents then approved as the collective expression of the civil society sector in the Plenary. Recognizing the peculiarity of such procedure, the documents we have analysed can well be considered as part of civil society convergences at WSIS.

18 For a complete explanation of the method and its application we refer to the appendix of the chapter. Here we recall that we conduct our analysis with simple words as well as with multi-words or sequences, which are defined as CTU. Complex textual units are used to a) increase the amount of information (textual units carry more information than simple terms) and b) to reduce the ambiguity of simple word-types (simple word-types are ambiguous because they are isolated from their context of usage). Graphs are built on the basis of correspondence analysis and we visualize some of the CTU that contribute to the definition of axes. Yet, our interpretation draws not only on the visualization of CTU in the graph but also on ulterior information concerning specific and exclusive CTUs. Specific CTUs: those relatively more used in a document or a phase in comparison to others, therefore relatively more important in those documents; and exclusive CTUs: those which have been exclusively used in a specific phase/set of documents.

19 The three official PrepComs (July 2002, February 2003, September 2003), the Informal meeting (Nov. 2002), the Intersessional meeting (July 2003), PrepCom3A (November 2003) and the Summit (December 2003.

20 For consideration about the choice and selection of governance CTUs, see Padovani & Tuzzi (2004).

21 Contributions to PrepCom1 were developed before the actual process started and therefore show a different language and focus mainly on value aspects such as: participation, democratic, consensus. Actors are identified in a generic form and barely mentioned (all countries, all citizens), the only element concerning the 'what' of governance is a single mention of 'regulatory framework'.

22 Once again we underline this aspect, since we consider the structure through which civil society has self-organized itself a meaningful space for dialogue among different realities. It should be recalled that a number of civil society actors maintained their own interest and language, as demonstrated by the number of civil society documents presented at the final Summit, which we have analysed elsewhere (Padovani & Tuzzi, 2004). Therefore we are aware that the documents we are taking into consideration are not fully representative of the entire presence of civil society at WSIS; nevertheless we consider them as the result of the most articulated collective effort of cooperation.

23 The closer a document appears to the origin of axes, the least specific its language in relation to other documents. Not all specific and exclusive CTUs appear in our graphs, in order to make the graph more readable.

24 This was actually the position expressed in the civil society statement presented at PrepCom3A, in November 2003, when the sector denounced the limits of the intergovernmental process and declared it was to write an alternative declaration.

25 Aspects of our methodology have been developed over time; therefore the two analyses cannot be subject to a direct comparison. Yet it is interesting to have some historical insights.

26 Bucharest, Beirut, Bavaro, Tokyo and Bamako represent two reports elaborated in the regional preparatory conferences held between PrepCom1 and PrepCom2 (a part from the Bamako meeting, which took place in May 2002). CCBI is the document elaborated by the Coordinating Committee of Business Interlocutors, private sector; Prep2 principles and Prep2 Action are, respectively, the draft documents for the *Declaration of Principles* and the *Plan of Action*; CSCG is the document elaborated by the Coordinating Group of Content and Theme, civil society sector; Samassekou stands for the 'non-paper' proposed by the President of the Preparatory process as a basis for negotiation of documents.

27 We refer to Padovani & Tuzzi (2004) for a complete analysis of all documents from the Summit.

28 As reported by Kaldor (2003: 96–97) when discussing the role of NGOs. See also Calabrese (2004).

29 The Civil Society Plenary document uses information_and_communication_societies 25 times and knowledge_societies 4 times (in an exclusive manner). This is its strongest statement: the idea of a plural reality, which should be respected in principle as well as through appropriate

wording (recurrent are: pluralistic, differences, linguistic_and_media_diversity) (Padovani & Tuzzi, 2004a).

30 Aggregative and integrative approaches have been elaborated by Messina (2003) building on new-institutional analysis, focusing on administrative styles in local governance. We here suggest that a similar interpretation could be adopted to describe the visions of governance that emerged from WSIS documents. This certainly needs further reflection, thus we see it as one of the many interesting starting point for future investigation.

References

Arts, B. 2003. *Non-state Actors in Global Governance. Three Faces of Power*. Working Paper Series (2003/4). München: Max Planck Gesellschaft.

Baylis, J., Smith, S. 1997. *The Globalization of World Politics. An introduction to international relations*. Oxford: Oxford University Press.

Bolasco, S. 1999. *Analisi multidimensionale dei dati*. Roma: Carocci.

Bolasco, S., Baiocchi, F., Morrone, A. 2000. TALTAC Versione 1.0. Roma: CISU.

Calabrese, A. 2004. 'The Promise for Civil Society: a Global Movement for Communication Rights', in *Continuum: Journal of Media and Society* 18(3): 317–329.

Cammaerts, B. 2004. 'Online Consultation, Civil Society and Governance: does it really make a difference?', paper presented at the IAMCR Conference in Porto Alegre, July 2004.

Cammaerts, B., Carpentier, N. 2004. 'The Unbearable Lightness of Full Participation in a Global Context: WSIS and Civil Society Participation', paper presented at the COST final conference on 'Transnational Communities', Rovaniemi, May 2004.

Cammaerts, B., Van Audenhove, L. 2003. *ICT-Usage's of Transnational Social Movements in the Networked Society: to organize, to mediate, to influence*, EMTEL2, key-deliverable, Amsterdam: ASCoR.

Carlsson, U. 2003. 'The Rise and Fall of NWICO. From a Vision of International regulation to a Reality of Multilevel Governance', in *Nordicom* 24(2): 31–67.

Castells, M. 2000. *The information Age, Economy, Society and Culture*. Oxford: Oxford University Press.

Castells, M. 2001. *The Internet Galaxy. Reflections on the Internet, Business and Society*. Oxford: Oxford University press.

Della Porta, D., Diani, M. 1997. *I movimenti sociali*. Roma: Nuova Italia Scientifica.

Della Porta, D., Kriesi. 1998. 'Movimenti sociali e globalizzazione', *Rivista Italiana di Scienza Politica* (3): 451–482.

Della Porta, D., Mosca, L. 2004. 'Global-net for global movements? A network of networks for a movement of movements', paper presented at Congress SISP, Padova, September 2004.

Diani, M. 2000. 'Networks as social movements: from metaphor to theory?', paper presented at the Conference 'Social Movement Analysis: the Network Perspective', Ross Priory.

Diani, M. 2001. 'Social Movement organizations vs. interest groups: a relational view', paper presented at ECPR General Conference 'Mobilization and participation', Canterbury.

Eschle, C. 2000. 'Engendering global democracy', paper presented at the IPSA congress, Quebec.

Falk, R. 1987. 'The global promise of social movement: explorations at the edge of time', *Alternatives* (XII): 173–196.

Featherstone, M. 1990. *Global Culture. Globalization, Nationalism and Modernity*. London: Sage Publications.

Frissen, V., Engels, L., Ponsioen, A. 2002. *Transnational civil society in the Networked Society. On the relationship between ICTs and the rise of a transnational civil society*, Study in the framework of TERRA 2000, International Institute of Infonomics.

Greenacre, M. J. 1984. *Theory and Application of Correspondence Analysis*. London: Academic Press.

Guidry, J., Kennedy, M., Zald, M. 2000. *Globalization and Social Movements. Culture, Power and the Transnational Public Sphere*. Ann Arbor: University of Michigan Press.

Hamelink, C. 2001. '(la governance) della comunicazione globale' in Padovani C. (ed.) *Comunicazione globale. democrazia, sovranità, culture*. Torino: UTET libreria.

Held, D. 1999. *Democrazia e ordine globale. Dallo stato moderno al governo cosmopolitico*. Trieste: Asterios Editore.

Held, D., Mcgrew, A. 2001. *Globalismo e antiglobalismo*. Bologna: Il Mulino.

Held, D., Mcgrew, A., Goldblatt, D., Perraton, J. 1999. *Global Transformations. Politics, Economics and Culture*. Cambridge: Polity Press.

Herman, E., Mcchesney, R. 1997. *The Global media. The New Missionaries of Global Capitalism*. London: Cassell.

Hewson, M., Sinclair, T. 1999. *Global Governance Theory*. Albany: State University of New York Press.

Kaldor, M. 2003. *Global Civil Society. An Answer to War*. Cambridge: Polity Press.

Keck, M., Sikkink, K. 1998. *Activists Beyond Borders*. Ithaca, NY: Cornell University Press.

Keck, M., Sikkink, K. 1999. 'Transnational Advocacy networks in the Movement Society', pp. 217–238 in Meyer D. & Tarrow S. (eds.) *The Social Movement Society. Contentious Politics for a New Century*. New York: Rowman & Littlefield.

Keck, M., Sikkink, K. 2000. 'Historical Precursors to Modern Transnational Social Movements and Networks', pp. 35–53 in J. A. Guidry, M.D. Kennedy & M. N. Zald (eds.), *Globalization and Social Movements. Culture, Power and the Transnational Public Sphere*. Ann Arbor: University of Michigan Press.

Klein, H. 2004. 'Understanding WSIS. An Institutional Analysis of the World Summit on the Information Society', paper available at www.IP3.gatech.edu.

Kleinwächter, W. 2001. *Global Governance in the Information Age*, Papers from the Centre for Internet Research, Aarhus, Denmark.

Kleinwächter, W. 2004. 'Beyond ICANN vs. ITU. How WSIS Tries to Enter the Territory of Internet Governance', *Gazette* 66 (3–4): 233–251.

Kooiman, J. 2003. *Governing as Governance*. London: Sage Publications.

Lebart, L., Salem, A., Berry, A. 1998. *Exploring Textual Data*. Dordrecht, Netherlands: Kluwer-Academic Pub.

Lee, M. 2003. 'A historical account of critical views on communication technologies in the context of NWICO and the MacBride Report', paper presented at the Euricom colloquium, Venice.

Lipshutz, R. 1992. 'Reconstructing world politics: the emergence of global civil society', *Millennium: Journal of International Studies* 21(3): 389–420.

Macbride, S. 1980. *Many voices, one world*. Paris: Unesco.

Mattelart, A. 2001. *Storia della società dell'informazione*. Torino: Piccola Biblioteca Einaudi.

Meyer, D., Tarrow, S. 1999. *The Social Movement Society. Contentious Politics for a New Century*. New York: Rowman & Littlefield.

Morrone, A. 1996. 'Temi generali e temi specifici dei programmi di governo attraverso le sequenze di discorso', pp. 351–69 in Villone M. and Zuliani A. (eds.) *L'attività dei governi della repubblica italiana (1947–1994)*. Bologna: Il Mulino.

Naughton, J. 2001. 'Contested space: the Internet and Global Civil Society', pp. 147–168 in

Anheier H., Glasius M., Kaldor M. (eds.) *Global Civil Society 2001 Yearbook*. Oxford: Oxford University Press.

Neveu, E. 2001. *I movimenti sociali*. Bologna: il Mulino.

Niggli, P. 2003. 'Une légitimité contestée', pp. 1–2 in *Global: Globalisation et politique Nord–Sud*.

Nordenstreng, K. 1999. 'The Context: Great Media Debate', pp. 235–268 in Vincent, R. Nordenstreng, K. & Traber M. (eds.) *Towards Equity in Global Communication. MacBride Update*. Cresskill, New Jersey: Hampton Press.

Nye, J., Donahue, J. 2000. *Governance in a globalizing world*. Washington: Brookings Institution Press.

Ó Siochrú, S. 2002. 'Civil society participation in the world summit on the information society: issues and principles', discussion paper for working group 1.

Ó Siochrú, S. 2004a. 'Civil Society Participation in the WSIS Process: Promises and Reality', *Continuum: Journal of Media and Society* 18(3): 330–344.

Ó Siochrú, S. 2004b. 'Will the Real WSIS Please Stand Up? The Historic Encounter of the "Information Society" and the "Communication Society"', *Gazette* 66(3–4): 203–224.

Ó Siochrú, S. 2005. 'Global Media Governance as a Potential Site of Civil-Society Intervention', in Robert Hackett and Yuezhi Zhao (eds.) *Democratizing Global Media: One World, Many Struggles*. Lanham, MD: Rowman & Littlefield Publishers.

Ó Siochrú, S., Girard, B. 2002. *Global Media Governance. A beginner's guide*. New York: Rowman & Littlefield.

Padovani, C. 2001. *Comunicazione Globale. Democrazia, sovranità, culture*, Torino: UTET.

Padovani, C. (ed.) 2004a. *Gazette*, special issue on the WSIS 'The World Summit on the Information Society. Setting the Communication agenda for the 21st century?', vol. 3–4, June 2004, Sage Publications.

Padovani, C., Nesti, G. 2003. 'Communication about Governance and the Governance of Communication. Looking at European policies for the Information Society', paper presented at the IAMCR Conference, Barcelona, July 2002.

Padovani, C., Nesti, G. 2003. 'La dimensione regionale nelle politiche dell'Unione Europea per la Società dell'Informazione', pp. 207–227 in Messina P. (ed.) *Sviluppo Locale e Spazio Europeo*. Roma: Carocci Editore.

Padovani, C., Tuzzi, A. 2003. 'Changing modes of participation and communication in a international political environment. Looking at the World Summit on the Information Society as a communicative process', paper presented in the Political Communication section of the IPSA Congress in Durban, July 2003.

Padovani, C., Tuzzi, A. 2004a. 'WSIS as a World of Words. Building a common vision of the information society?' *Continuum: Journal of Media and Society* 18(3): 360–379.

Padovani, C., Tuzzi, A. 2004b. Debating communication imbalances: from the MacBride Report to the World Summit on the Information Society. An application of lexical-content analysis for a critical investigation of historical legacies, available at http://www.dssp.scipol.unipd.it/sisp2004.

Pianta, M. 2001. 'Parallel Summits of Global Civil Society', pp. 169–194 in *Global Civil Society Yearbook 2001*. London: The Centre for the Study of Global Governance, London School of Economics.

Raboy, M. 2004. 'WSIS as Political Space in Global Media Governance', *Continuum: Journal of Media and Society* 18(3): 347–361.

Raboy, M., Landry, N. 2004. 'La communication au coeur de la gouvernance globale. Enjeux et perspectives de la société civile au Sommet Mondial sur la Société de l'Information ', available at www.lrpc.umontreal.ca/smsirapport.pdf.

Rosenau, J. 1992. *Governance without Government. Order and Change in World Politics*. Cambridge & NY: Cambridge University Press.

Rosenau, J. 1999. 'Toward an Ontology for Global Governance', pp. 287–302 in Hewson M. & Sinclair T. (eds.) *Global Governance Theory*. Albany: State University of New York Press.

Rosenau, J., Singh P. 2003. *Information Technology and Global Politics. The Changing Scope of Power and Governance*. Albany: State University of New York Press.

Smith, J., Chatfield, C. Pagnucco, R. (eds.) 1997. *Transnational Social Movements and Global Politics. Solidarity Beyond the State*. Syracuse, NY: Syracuse University Press.

Tarrow, D. 1999. *Power in Movement. Social Movements, Collective Action and Politics*. Cambridge: Cambridge University Press.

Tehranian, M. 1999. *Global Communication and World Politics. Domination, Development and Discourse*. Boulder: Lynne Rienner.

Traber, M., Nordenstreng, K. 1992. *Few Voices, Many Worlds. Towards a Media Reform Movement*. London: World Association for Christian Communication.

Tuzzi, A. 2003. *L'analisi del contenuto*. Roma: Carocci.

UN General Assembly. 2001. *Resolution* 56/183.

UN Development Programme. 1999. Rapporto sullo Sviluppo Umano n. 10 'La globalizzazione'. Torino: Rosenberg & Sellier.

Van Audenhove, L., Cammaerts, B., Frissen, V., Engels, l., Ponsioen, A. 2002. *Transnational civil society in the Networked Society. On the relationship between ICTs and the rise of a transnational civil society*, Study in the framework of TERRA 2000, International Institute of Infonomics.

Wallerstein, I. 1990. 'Antisystemic movements: history and dilemmas', pp. 13–53 in S. Amin et al. (eds.) *Transforming the Revolution: Social Movements and the World System*. New York: Monthly Review Press.

3: Civil Society's Involvement in the WSIS Process: Drafting the Alter-Agenda

DIVINA FRAU-MEIGS*

This analysis of the WSIS process bears on three sets of results. One is the increased legitimacy of the role of NGOs within the ranks of other civil society actors, the other is the emergence of an alternative paradigm based on the cognitive revolution within the process itself, the last is the renewed place of research in the development of possible 'knowledge societies' (as an alternative to the unique paradigm of 'information society').

Civil society and the contents and themes drafters decided not to present an alter-agenda from scratch, contrary to other actors, like those involved in the CRIS campaign for instance. This was due partly to several factors: so many NGOs were involved that none in particular could claim the legitimacy to set the agenda; the necessary process of consensus-building needed inner negotiation; the very structure of the summit put civil society in a reactive rather than proactive situation, at its inception at least. Participants had to undergo their own process of self-knowledge and compromise, with very different backgrounds (some in research, others in activism, others in volunteer work). Also, after some early debates, the decision was taken not to be perceived as 'anti' but as 'alter', very much in parallel with the alter-globalization movement, which induced civil society to try and address the same questions as raised by the nation-states.

Consequently, civil society did not set the agenda, it assessed the proposals emanating from the nation-states and answered them step by step, very reactively, in some cases, especially in PrepCom1 and PrepCom2, within a day of the publication of the official documents. By responding fast and to the point, with professionalism and expertise, NGOs earned the respect of initially hostile or sceptical nation-states. As a result of this capacity for arguing and for implementing a soft-yet-firm civil disobedience, which did not balk at intense lobbying with the official representatives of supportive nation-states, some gains were obtained. Swiss researchers of the Institut universitaire d'études du développement de Genève even found that more than 60% of civil society's language was adopted in the official documents by the end of PrepCom2 (Institut universitaire d'études du développement de Genève, 2004).

Having thus prefaced the action of NGOs and civil society at large, I would like to examine some aspects of process and substance, to analyze the limitations of this strategy and also its forays into progress. How has process affected the drafting of the agenda? What substance was embedded in the final civil society document? How does it relate to the official documents? What does it bode beyond Tunis 2005?

Process

Limitations

The strongest limitation may well be the structure adopted for the summit, meaning the choice to opt for families and caucuses as reference groups. The families were decided through a top-down process, and crippled the possibility to draft an agenda that would address transversal issues, or issues not addressed by the nation states (public domain and e-commons for instance). The coordination was assumed by the Secretariat of the ITU, which was frustrating, as civil society would have preferred to coordinate itself. The suspicion was that ITU might manipulate the outcome by manipulating the coordination. Also, the Bureau family structure which involved family members more than caucus members was only partly mitigated by the composition of the Content and Themes group, based on a broader range of working groups and caucuses. This dual structure created some tension at times and forced a kind of self-selection.

Tensions arose because the families were perceived as representing traditional constituencies, modelled on the governments' framework, whereas caucuses were built on single issues and around areas of interest (human rights, gender, Internet governance, indigenous people, intellectual properties issues, etc.). Families have also been perceived as the single point of access for civil society but people tend to forget that the main organ of civil society is its plenary. The Bureau cannot make content-related decisions, only procedural and formal ones. Self-selection was induced by the number of meetings and their overlapping schedules; people who were in a capacity to partake in a number of issues, or who felt that there should be a systemic, global approach, felt frustrated because they had to make often mutually exclusive choices. This was the case for the International Association for Media and Communication Research (IAMCR), for instance, which had legitimacy both in the media family and the education family, but finally made its contributions mostly in the education family. We were thus capable of having the word 'research' added in the documents, in a prominent view. However, via the drafting team, we were also capable to make sure that the media ideas we promoted (including community media, public service media, etc.) were maintained in the final document.

Another nagging limitation was the language issue, with very high frustration levels due to the majority of documents being in English. This was particularly true for Spanish-speaking people and for French-speaking people from Africa, whose lists were among the most active. It had the result of creating a sort of self-selection of the people finally involved in the drafting of the final document, people like me, relatively familiar with three or more languages, which may have skewed the legitimacy of the drafters and their capacity to claim they represent a larger constituency. It relied more on good will, motivation, availability and language skills than on formal structures for representation. If people had been able to write their own claims in their own language, less would have been watered down. Those who were left, were more adept to mimic the kind of UN-ese language that passes off as English in official circles. Original or 'appropriate language', often asked of us in the drafting committee, was then more difficult to achieve. At the same time, it made it probably more acceptable to the authorities and more available for adoption by officials.

Progressive communities, coming as they were from many different backgrounds and constituencies, were not unified, and still are not. The divergence appears in the levels of involvement, in the priorities to be given to the agenda, in the choice of headings and banners for the final documents, etc. So civil society's strategy needs more integration, more cohesion, to reach a final integrated agenda. However, some lines of strength were identified, in like-minded groups, among which two have to be underlined: cities and municipalities on the one hand, small businesses on the other hand. They seem to coalesce on hands-on, community-oriented approaches, and are very pragmatic about the means to achieve their goals at the local level while being very outspoken and organized about their needs. It is in their direction that coalitions have to be constructed; they have a 'natural' capacity to develop viable multiple stakeholder structures.

Progress and Forays: NGOs as a Specific Actor within Civil Society

The role of NGOs has been essential, though it was not accepted as legitimate in the beginning of the WSIS process (and remains under question for the second phase of the process, in Tunis). Their status was questioned, especially when compared with more organized actors of civil society as defined by the United Nations (which include municipalities, trade unions, etc.). Older collaborators with the nation-states like the corporations of the private sector also objected. Doubts were expressed about the capacity of NGOs to organize, to master different approaches and appreciate the stakes, to resist the temptation of secession or withdrawal from the process altogether, and to gather the sufficient resources to establish a real presence. Yet on the very spots of the negotiations, in Geneva and Paris, NGOs found themselves in the position of direct interlocutors of the nation-states.

NGOs were in fact able to test the information–communication paradigm as a reciprocal space and a temporary zone of shared knowledge and collaborative work. They were able to use the structuring capacities of their networks to consult with their base and reach over large distances, in spite of some shortcomings, mostly due to language barriers. They were slightly overwhelmed by the final stages of the WSIS process, which required a significant presence in Geneva, but the by-then familiar use of the list, their knowledge of their reciprocal positions and the general guidelines and benchmarks they had adopted, allowed them to bypass this difficulty.

So, the Internet technology-adopted and adapted as a relational collaborative space by NGOs – proved useful for their goals. The capacity to mobilize real people through virtual communication, to create interaction, was made possible by certain congruence between cause, medium and network. It allowed NGOs to protest on the spot, to lodge complaints and requirements, and to participate in a constructive way, though they could only claim to be a non-representative but operational sample of global public opinion. This was not *per se* an experience in direct participatory democracy. It was rather an experience that showed that the Internet could work as a delocalized public forum, though the nation-states would like it to remain the common carrier it currently is. In fact, the Internet allowed NGOs to circumvent some of the hurdles of locating forums in traditional national capitals or international venue sites like Geneva, that are more easily controllable by traditional political bodies.

The technology helped NGOs in their capacity to organize civil society in a relatively coherent way. NGOs – real networks in their own right – used the Internet to enhance their capacity for mobilization, exchange, debate, as well as evaluation of the different steps of the WSIS process. It has allowed them to evaluate the role of communication within the political process, and to locate it between mediation and mediatization. In some cases, paradoxically, it has also allowed NGOs to protect the sovereignty of the states against their own tunnel vision, their tendency to accept interpretations of national sovereignty as interpreted by their inheritors, the transnational corporate world.

One problem remains however: the nagging feeling that NGOs tend to represent less a global public opinion than segments of the global population that are sensitive to issues of dependency and access, even if they belong to the middle class and are part of an intellectual elite (in ways trade unions or peasant coalitions are not). This was apparent in the functioning of the drafting group, more on the basis of coalitional tacit 'trust' than formal mandates from their respective NGOs or even their families and caucuses. It was probably reinforced by having to take position on an agenda mostly set by the nation-states and by writing on single issues, a practice that cannot produce the maximal level of implication and endorsement. Single issues also imply underlying issue networks and issue participants. They tend to blur the global picture and the general interest. This confirms, if need be, that nobody can expect the technology alone to create participation and direct democracy. The political implication of citizens is of the essence and those motivations are not technological, they are social.

One failure is worth pondering over. NGOs have failed to get the attention of the general media outside the WSIS process, before and during the summit, even if they have used effective media repertoires and strategies and communication skills within civil society. This can be explained partly because it was not part of the media agenda to deal with a subject so close to the quick, partly because NGOs remained guarded from journalists and other people who risk to implement their views to serve their own agenda and bias. They have learnt to avert the increasing tendency of institutional media to represent views offering progressive proposals for change in negative ways. And yet they do need to broadcast their ideas in the mainstream of national populations.

Substance (On Education and Research)

Limitations

Education has been more advantageously dealt with than research. In the official view of the nation-states, when it is mentioned, it relates to R&D, in the industrial perspective of applied and hard sciences, basically connected to utilitarian technological advances and product development. The soft sciences have been consistently neglected in the process. This can be explained partly because they have no apparent link with information technologies, partly because they are openly critical of the all-technological approach and favour a 'social uses' approach for technology to meet a local demand and offer a solution to real problems. Most nation-states coming to the summit were only interested in acquiring the latest technology (IPV6), with a view of rationalizing governmental functions only (including surveillance and monitoring of citizens). It is still the main purpose of the next stage, whose official focus is on financing the global infrastructure of Information Society and deciding who runs the Internet.

Social sciences are also critical of buzzwords and they have cast doubts about the phrase 'Information Society'. The civil society document reflects this careful weighing of the meaning of the words, by systematically replacing Information by Knowledge, by associating Information with Communication, by adding an 's' to Society, thus acknowledging the diversity of cultures.

More disturbingly, social sciences research underlines the difficulties of articulating information, expertise and know-how at the local level. It casts doubts on the facility to realize fast the full potential of digital dynamics for the populations in need or marginalized, worldwide. While extolling ICTs' capacities for the empowerment of individuals and communities alike, it also underlines uncertainties about the social outcomes, the real needs, the failures and the risks. It asserts that Knowledge Societies will fail, if no self-supporting system for culturally appropriate learning and research practices is established, in these areas for which the information and communication technologies hold out, paradoxically, the greatest promise for material and humanistic gains.

Here too the various sub-groups representing civil society were not unified on the meaning of research for education and its connection with public domain issues and open access. The divergence appears mostly in the priorities to be set at the top of the agenda. Some wanted to focus on basic literacy (not even digital literacy), others privileged training for jobs and labour, others wanted to push infrastructure and access, etc. Besides it was never clear if there was a total convergence between hard sciences and soft sciences on the issues at stake. Though the civil society declaration managed to integrate the gist of some of the documents presented in other events leading to WSIS (like the Budapest Open Access Initiative, Berlin Declaration, Creative Commons), the alternative agenda for research and education appears as watered down and scattered across the document; the official documents show the same indecision, which points to the fact that the NGOs' strategy needs more integration, more cohesion.

As a result of these limitations, the civil society declaration and the official documents alike provide little or no attention to the means, no financial proposal (no real mention about who finances and how: no clear positioning on the African suggestion for a Solidarity fund, no precise modalities for oversight and monitoring of the *Plan of Action*, etc.). Everything has remained too abstract, especially at the *Plan of Action* level. This lack may explain the two issues that have been singled out for further discussion, Internet governance and financing, but they are problematic as such and many feel frustrated because they are convinced all the other points on the agenda need to be attended to. This may partly explain the relatively low level of endorsement of the civil society document at this point. But it seems to reflect the progress and consensus that could be reached under difficult conditions of time, space, connection, language, etc.

Progress and Forays

A general consensus however seems to have formed around education. The official document and the civil society document both extol it as a principle and as a need. Because it seems the most democratically acceptable for all, there has been no heated debate over it, contrary to other issues like human rights or intellectual property rights. However the two documents are in fact divided over a common value. While the nation-states tend to privilege education for the creation of an efficient labour force, civil society sees education and literacy as a means to build lifelong autonomy and collaborative exchanges. Civil society considers education on a continuum of knowledge, consistently connecting it to related issues of access, capacity-building, community-based solutions, public domain commons, linguistic diversity and pluralistic approaches to cultures.

During the WSIS process civil society has slowly but surely been able to reassess the modifications introduced by globalization and by technological possibilities for empowerment. As a result it has embedded in its declaration an alternative model for research and technology, different from the traditional R&D model of the industrial age. This industrial model, which served the Western world for two hundred years, relied on stable scientific disciplines, with their borders clearly marked, with their maps of knowledge and their hierarchy of content, with their strict selection of scientists and engineers at the entrance-level, with their own sets of evaluations, standards and intellectual property laws. This inherited model, which has accompanied the spread of nationalism, tends to favour some European countries, the United States and Japan, with a balance of power tilting towards international corporations emanating from these very nation-states.

What the process has also revealed is the cultural conflict, even within the industrial model. Some members of the world of computer science and research have joined the ranks of civil disobedience, questioning the monopolistic practices of multinational corporations and their claim that their interests are to be equated with the economic interests of the whole world. As a result, it seems that expressions of general interest are emerging from the margins ... and from within, which is what we have been witnessing in the emergence of this embedded alter-agenda.

The Alternative Agenda

Scattered in its various sections and sub-sections, the civil society document offers an alternative model of open 'R&C' (Research and Collaboration), part and parcel of the new informational model, whose various component have only recently gelled in a coherent whole. Its key elements point to the sustainable spread of prosperity beyond material goods and their market reproduction, to include knowledge and a better functioning of the world society. It purports that to be up to the potential of ICTs (Internet and beyond), there is the need to elaborate a complex understanding of how our cognitive and semiotic resources have elaborated media uses and regulations within a given culture (Merlin, 1991; Norman, 1993; Kunstler, 1996). It supports the idea that the scale for primary human associations needs to reinvent the local *community of place* (Quartz & Sejnowski, 2000: 274).

Embedded in the civil society documents, there is a cognitive revolution at work that predicates a different view of human nature. This is basic to all real change, as exemplified by the prior revolution of that sort, the Enlightenment revolution. The view of human nature derived from the eighteenth and nineteenth centuries worked under the assumption of self-interest and the notion that man is a wolf for man, the need for a coercive state and the regulation by the market (Mansbridge, 1990). It led to predicating the legitimacy of media on the notion of freedom of expression. Three centuries later, the knowledge about human nature has drastically modified this picture; so has our environment (Clark, 1997; Tomasello, 1999; Harrison & Huntington, 2000). This new knowledge has elicited a view of human nature as collaborative, expressing itself in an open-ended process of distributed intelligence and exchange with the environment (Salomon 1993). Plasticity, portability, responsiveness, connectedness, such are the new keywords attached to these cognitive advances.

This view extends the reach of freedom of expression into the realm of social capital and truly situated knowledge societies; it mitigates the view according to which human nature is individualistic, solely driven by instincts that need to be curbed by the state. It encourages the recourse to forces of civil society for participation in the regulation of media, and especially the Internet, as a tool for renewed connectedness with a common purpose. It has the potential to lay the grounds for a new political theory predicated on cognition and using the distributed intelligence of the Internet network as its media of choice conveyance (Quéau, 2000). Though it has not yet produced visible changes in the political and legal domains, its challenging views are creating a situation of instability and uncertainty in culture, very perceptible during the WSIS process.

This view is gaining importance because the Internet is perceived as having unacceptable real world effects on people. It is seen as a medium for terrorism, cybercrime, spam, all issues that have appeared on the WSIS agenda and have displaced the access and rights issues. There is an increasing overlap between real world decision-makers and Internet decision-makers as the founding fathers of the media give way to more ordinary users and developers. In spite of Lawrence Lessig's much touted phrase that in cyberspace 'code is law' (Lessig, 1999: 6), the notion that technicians should decide of norms without accountability is being challenged by the call for anchorage in national laws, if not international ones.

So two models are at work in the process, in relation to the regulation of ICTs. There is on the one hand an explicit *information-provider* model that relates NTICs to any commercial model, likening them to a raw resource, to be exploited for economic development; it recalls the invisible hand of capitalism and individualistic greed. In such a view, economies of scale still are one of the guiding principles of the design of social arrangements. But more importantly there is also an Open Source model, with a technology attached to it (open-code software). It also refers to an implicit societal organization, that of the Creative Commons. This approach is based on public domain preservation and enhancement, to be achieved by convincing content producers to be active participants in the open-access paradigm of knowledge, along the lines delineated in a variety of documents and initiatives (Budapest Open Access Initiative, Berlin Declaration, Creative Commons, Open Courseware Initiative, etc.). Trying to promote participation and transmission, it is the only approach predicated on a cognitive view of human nature as collaborative, responsive and involved in a distributed, sustainable exchange of intelligence. Hence the fact that the civil society documents underline the need for community-based, self-supporting systems, with in-built maintenance programs and upgrading capacities. They also call for the free flow of knowledge, the public domain preservation, the active participation of content producers in the open-access paradigm of knowledge, along the lines delineated in a variety of documents and initiatives like those mentioned above. (Frau-Meigs, 2005).

The co-presence of these two models suggests the possibility of a bifurcation of cultures within the Internet environment, to accommodate their diverging trends: on the one hand a protraction of the media commons culture, on the other hand a protraction of the commercial market culture. Yet recent initiatives such as the BBC's Creative Archive, which allows users to download and modify digital clips of BBC television, illustrate the roads that may be taken, the data mining of archival repositories being done on an open-access perspective. Another initiative the Open Courseware initiative has also emerged (supported by MIT, ParisTech, Moscow University...), proposing open access to their engineering courses on the Internet.

These proposals make sense with the world picture. Worldwide there is a growing distrust in federal government service delivery and a sense of disenfranchisement. A variety of societal movements are promoting ethnic identities, devolution of state rights, and community building at local levels. They express the need for human connectedness and the feeling that global media have not provided the appropriate scale for human interaction (Castells, 1997). Though flawed because of its focus on the private rights of the individuals to the detriment of a balance between private and public needs and spaces, for a common purpose, this perspective may bring some political changes and modify people's perception of their use of the technology.

At this stage, the end of part I of the summit and the beginning of part II, it seems clear that civil society has been able to plant the seeds for alternative and competing views on research, education and technology within the official documents and within the minds of government officials. It has acted as a wedge actor, with a certain amount of leverage, due to the tensions within the old model. Some hybridization process is at work, between traditional, industrial and national forms of knowledge production, not yet obsolete and still quite efficient, and new forms that appear as viable international alternative models for the production and exchange of knowledge. Governments may find themselves as arbitrators between the two, trying to keep a balance between the need for public connectedness and the drive for private business. Potential changes, for the future, will come from this dialogue, at times painful, at times fruitful, between the corporate sector, the governmental world and the civil society actors. In this tripartite collaboration, NGOs have surprised by their force of proposal and their capacity to stay into the process. Some of their language and their claims, already appropriated by nation-states, are probably going to be institutionalized, hopefully towards more cultural pluralism and a more diversified use of media and technologies for the building of knowledge societies.

Another kind of hybridization is also appearing between promoters of direct participation and promoters of political representation. Some actors have weakened, like trade unions and parties, but others have gained strength; NGOs, for instance, to the point that some governments, like the US, have felt the need to create an NGOwatch (via the American Enterprise Institute), to monitor the lobbying efforts of these relatively new actors. These trends show the need to strike a new balance between the power of civil society actors, the nation-states and the private sector. Hence, in spite of current resistance from the corporate world, there will probably be a shift in favour of a new balance of intellectual properties as a common ground for individual creators to protect their works and for civil society users to benefit rapidly from their contributions. The ingenuity of solutions that need to be found is also exemplified in the movements for digital checks and balances and for the transfer of Internet governance, away from proprietary private hands.

The new balance will strike a modus vivendi between political mediation and technological mediatization, and some actors will suffer more than others. The NGOs that will be most capable of federating not simply around single issues but around general interest issues, in association with related social movements, will be the most likely to push their vision and foster social change. It is essential that these tendencies do not lead to the privatization of public space or to the erasure of global public issues. NGOs must stand watch, as the new tripartite governance in the making cannot simply model itself on a corporate organization of functions, powers and knowledge production. More political and social awareness needs to be produced at the level of the WSIS in the years to come, even beyond stage II. The process is far from being finished; its best institutional use so far has been the possibility for NGOs and researchers to test the strength of their ideas, in the interest of the broadest possible civil society.

The Role of Researchers in NGOS and the WSIS Process

Managing a Bi-Polar Situation

The role of the research community, taking into account the soft and hard scientists and also the input of some socially aware and responsible computer researchers and professionals, has consisted in being providers of complex explanations and long-term understanding of competing views of the technological world. This role is not going to diminish as our societies become increasingly global and as the need for systems of global conflict resolution and for shared knowledge, the so-called 'world governance', is expanding. The researchers were able to help NGOs and other civil society actors to articulate their views and to organize their participation, more painfully probably than the private sector and other stakeholders, because of their own self-imposed double bind of respecting pluralistic views and yet couching them in an all-encompassing language acceptable by all. Paradoxically also, if a general survey was made of those most implicated in NGOs and Content and Themes, it probably would find a lot of people trained in the social sciences or doing research in a social sciences perspective, with a majority of women.

As a result of the WSIS process, the debate within the research community has been re-launched about its capacity to react fast and to make a difference. Researchers have come to the realization that they must keep working at a double task: maintaining a cool distance from events and yet providing some compelling piece of thought, to feed to the NGOs and to governments. They have the responsibility of making sure their informed point of view penetrates the global public space, so that their community remains engaged in the national and international debates. They have already taken the risk of engaging in proposals of models for action, in open procedures that have to be constructive and not just critical of institutional and economic logics.

The current moment however shows a bi-polar situation for researchers: they work within institutions inherited from the industrial age paradigm, which endures in spite of increasing malaise, and they are activists in instances that are very fragile as all NGOs are. Trying to rethink their practices and modes of production of knowledge, they must take full advantage of the opportunities offered by the information and communication paradigm. They stand in between two worlds, between the weightiness of their scientific real-life activities and the lightness of their digital on-line activities that give visibility to their alternative views. Uncomfortable as they may seem, both stances are necessary so as not to produce 'more of the same', so as to re-invent the profession and its modes of exchange and knowledge. This remains their main social function and justification in this global process (Frau-Meigs, 2005).

Keeping Watch on the Future: Implications for WSIS and Beyond

Researchers also have the benefit of hindsight that they can apply to foresight on how ICTs may evolve. At the moment, the risk is for both 'enclosure' and 'broadcastization' of the Internet. The Internet is being turned into a media rather than a network of networks. Its novelty as an interactive communications tool is being partly 'naturalized' or 'normalized' by society and societal uses. The latest commercial trends show that there is a tendency to assimilate it to other existing mass media. Its development is closely co-related with other media businesses and as a result it is increasingly used for a variety of complementary services anchored in territorial grounds. These numerous commercial intermediaries aim at an enclosure of the open-ended system: they only care to give access to the services they have a stake in, often connected to other media entertainment and information processing strategies. This surreptitious enclosure is supported by research on the uses of the Internet. It confirms that a majority of users explore little beyond the sites and portals offered by the major providers. This is a real limitation to the end-user, and the citizen at large, as the commercial architecture of the network allows service providers both to trace and monitor usage and to constrain freedom of navigation.

The post 9/11 context also illustrates this tendency, as the American military are pushing for more surveillance of the Internet, to buttress their anti-terrorist policies and related cyber wars. Such steps have significantly and permanently altered any American goodwill to modify a national sovereignty and integrity position: *Realpolitik* has made a singular return with the Bush doctrine (Lafeber 2002: 543–556), whose principle is 'what is good for the US is good for the world', to justify isolationism and unilateralism. These events have been concomitant with the end of the first expansion phase of the Internet and the necessary legal stabilization that the industry calls for. They have made the virtual world contingent with the real world, dramatically so. They may have damaged durably the generous impulse of collaborative exchanges that was at the foundation of the World Wide Web, founded around the researchers' needs for collaborative solutions to specific questions they had.

Within the US, the pressure is high from intelligence-gathering agencies like the FBI, NSA and other military entities, to proceed to a closure of the open-ended system, as has been the case with other media in the past. The Internet Engineering Task Force is still in a capacity to resist and maintain some openness in the system, but for how long? It is also under pressure from the industry, which would like to use its expertise for strictly corporate purposes, as in the case of Microsoft. In fact, other industrial sectors have their stake in the closure of the system, which will allow a clearer way of defining costs, billings, returns on investments, etc. They do not welcome the Open Source and Open Software initiatives that would make this data mining more largely accessible, collaborative and free ... as the recent conflicts about intellectual property rights have shown.

As a result, the cognitive model and its keywords of citizen direct participation (lifelong training, sustainability, attention to indigenous cultures, not to mention cultural diversity, open source and open access) may become the reserve of a limited number of diehard research amateurs on the one hand, and of impoverished indigenous minorities on the other, both relegated to the local spectrum, which is perceived as neither commercially viable nor strategically threatening. As with past 'new' media has since become old, they may continue to do their own tinkering, making up micro-communities of radio hams, CB users and, now, potentially, Internet hackers. In fact, conflict may arise between the two extremes of democratic tension, the amateurs confronting the military while the middle forces (corporations, operators and the government) exploit their antagonism. When amateurs gleefully show up the weaknesses of a system or claim greater flexibility through spectacular operations such as sending viruses onto the sites of government agencies or major corporations, the military demand more security and more surveillance, which is renegotiated by the government and the corporation without public consultation. These are recurrent arguments in the history of media, applied to radio and television earlier on, visibly at work with ICTs now (Frau-Meigs, 2001).

The consequences for the WSIS process, imperfect as it is, may be damaging if not carefully monitored because it endangers the tripartite involvement of civil society, private sector and nation-states. Civil society might become at best the equivalent of the *tiers-état* of France before the Revolution, when in fact it should be considered on a more equal footing; also civil society seems to be relegated to the role of community-building only, as if it had no competence in other domains. Though the multi-stakeholder approach was made mandatory in the WSIS process by UN Resolution 56/183 (December 2001), the concept is not clearly described, even in the official documents that were the outcome of the first phase of the summit (Geneva, 2003). Article 49, while asserting the need for a plural approach to Internet governance is unclear about each actor's respective functions and accountabilities; it shows the hesitancy between several models for media regulation, with a tendency to underplay the role of traditional media. It recognizes that:

a) Policy authority for Internet-related public policy issues is the sovereign right of States. They have rights and responsibilities for international Internet-related public policy issues; b) the private sector has had and should continue to have an important role in the development of the Internet, both in the technical and economic fields; c) civil society has also played an important role on Internet matters, especially at community level, and should continue to play such a role; d) intergovernmental organizations have had and should continue to have a facilitating role in the coordination of Internet-related public policy issues; e) international organizations have also had and should continue to have an important role in the development of internet-related technical standards and relevant policies. (Declaration, article 49 section 6)

Different constituencies are recognized but they still have to stake out their territory, their legitimacy and their grounds for accountability. The inclusion of the private sector and civil society, i.e. non-governmental stakeholders, is not yet completely integrated in the mechanism. It indicates that a trilateral model of global governance is still in the making as co-regulatory policies are difficult to envision within a framework of national sovereignties. The nation-states, under pressure of operators and corporations, are mostly concerned with a narrow approach and technical standards. Policy-makers find it difficult to adopt a bottom-up strategy that would relinquish part of their power to a larger number of stakeholders.

The outcome of the first phase of the WSIS explicitly calls for a media-specific international Internet governance (*Plan of Action*, articles 13B, 13C, 13D under section 'enabling environment'). So the functions of the different stakeholders will be defined as task-specific and they may remain narrow and technical, giving an edge to the private sector and the telcos. A larger understanding of ICTs and of Information Society will have to emanate from other processes, more political and legal than technical. The compromise, negotiated, solution seems to be the inclusion of a fourth actor, Non-Governmental Institutions (NGIs), which gains control over the others and is not without implications for researchers, NGOs and civil society. ICANN and the International Telecommunication Union (ITU) were present from the start but UNESCO, long discarded, made a significant comeback in the drafting of the final documents (more than half the points of the action plan are under its constituency) and in the events taking place during the WSIS Summit itself (Geneva 2003). ICANN has been more and more controlled by its Governmental Advisory Committee, the consultative body of nation-states that is part of its framework; it has agreed to respect the national legal environments of each country. ITU represents the technical interests of telcos; it is controlled by an industry–government partnership. UNESCO provides a broader, cognitive view on culture; it has adopted the open code software, which brings it close to the Open Source and Creative Commons model, all the more so if it is combined with the cultural diversity model, whose regulation is under its mandate. So if ICANN and ITU tend to be strictly technical, UNESCO provides for a cultural alternative.

Currently, on a global scale, the only model that takes care of the local needs of communities and tries to translate them into an international law is the cultural diversity model placed under the auspices of UNESCO (Frau-Meigs, 2002: 3–17). It is the only model that incorporates traditional and new media, but also all sorts of cultural goods and services into an international framework and as such it has to be observed carefully. It implies that the state is the intermediary link that fosters community-building and maintains cultural pluralism within its borders, provided it nurtures the paradigms and values of its diverse constituencies. It sets the nation-state as a wedge intermediary, facilitating the arbitration of interests between the local and the global. It acknowledges the fact that it is difficult to argue for a single, unique model of governance while acknowledging the human need for situated communication and distributed cognition.

Ideally, an enhanced communications process should emerge from the cultural diversity model, allowing territorially based communities to protect their vital interests and let it be known to Internet participants when their online actions threaten them; conversely, online participants should be able to inform offline communities when they feel that their online rights and freedoms are being unduly touched upon. What needs to be internationally devised is a system of accountability and inter-operability, no more no less. Interestingly, the regulatory emergence of this model has probably prompted the US to re-incorporate UNESCO, in an attempt to thwart it, as it is a reminder of past WTO disputes on the topic of cultural exception. Interestingly also, it has also been relayed within the UN framework of WSIS, as the *Declaration of Principles* explicitly supports UNESCO' *Universal Declaration on Cultural Diversity* (article 52, section 8). UNESCO plans to bring to the second phase, in Tunis 2005, a full convention on cultural diversity, making it into a right, to be added to the other human rights (with the attendant sanctions attached).

None of the NGIs can represent civil society's plea for a more decentralized bottom-up solution. ICANN and ITU seem too much tilted toward private commercial targets and American-dominated interests. UNESCO seems too much the realm of nation-states sovereignty, with little bottom-up capacity, in spite of the increased synergy it tries to develop with NGOs around the world. At the global level, it seems that the tripartite, multi-stakeholder approach will have difficulties in getting under way, as there is at the moment little consensus about the stakes, the functions, the respective needs of the various actors. The governments speak with many voices, though they are in agreement about their sovereignty as states; the private and commercial entities are also divided, though they share a liberal view of the marketplace; civil society has not reached a consensus either, though it pleas for an open program and process, guided by transparency and a bottom-up approach. But the process itself is making a creative use of collective visions; alternative paradigms and metaphors for action are being circulated widely. Without intending it, the WSIS process is functioning as the largest consultation offline and online that has yet been undertaken on the management of media resources. This in itself is a positive sign that a measure of change is under way.

Note

* *In spite of my institutional involvement in the WSIS process, as vice-president of the International Association for Media and Communication Research (IAMCR), as focal point for the 'education, research and academia' family and as part of the 'content and themes' drafting committee, this paper reflects only my personal views.*

References

Plan of Action, Article 13b, 13c and 13d under section 'enabling environment', document WSIS-03/GENEVA/DOC/5-E, 12 December 2003, available at www.itu.int/wsis/index.html.

Castells, M. 1997. *The Power of Identity*. Oxford: Blackwell Publishers.

Clark, A. 1997. *Being there: putting brain, body and world together again*. Cambridge: MIT press.

Declaration of Principles, article 49 under section 6 'Enabling environment', document WSIS-03/GENEVA/DOC/4-E, 12 December 2003, available at www.itu.int/wsis/index.html.

Declaration of Principles, article 52 under section 8 'Cultural diversity and identity, linguistic diversity and local content'; document WSIS-03/GENEVA/DOC/4-E, 12 December 2003, available at www.itu.int/wsis/index.html.

Frau-Meigs, D. 1998. 'Cybersex, censorship and the State(s): pornographic and legal discourses' *Journal of International Communication*, special issue 'human rights', 5 (1–2): 211–27.

Frau-Meigs, D. 2001. *Médiamorphoses américaines dans un espace unique au monde*. Paris: Economica.

Frau-Meigs, D. 2002. 'La excepcion cultural en una problematica intercultural', *Quaderns del CAC* (14): 3–17. In English, French, Spanish and Catalan, available at www.audiovisualcat.net.

Frau-Meigs, D. 2005. 'On Research and the Role of NGOs in the WSIS Process', The World Summit in Reflection: a deliberative dialog on WSIS (special issue). *The journal of Information Technologies and International Development* (forthcoming).

Harrison, L. E., Huntington, S. P. 2000. *Culture Matters: How Values Shape Human Progress*. NY: Basic Books.

Institut universitaire d'études du développemnent de Genève. 2004. 'Sommet mondial sur la société de l'information, phase I, Genève, 10–12 décembre 2003', *Annuaire suisse de politique de développement* 23(1): 147–158.

Kunstler, J. 1996. *Home from Nowhere: Remaking Our Everyday World for the Twenty-first Century*. NY: Simon and Schuster.

Lafeber, W. 2002. 'The Bush Doctrine', *Diplomatic History* 26 (4): 543–56.

Lessig, L. 1999. *Code and Other Laws of Cyberspace*. New York: Basic Books.

Mansbridge, J. (ed.) 1990. *Beyond Self-interest*. Chicago: Chicago UP.

Merlin, D. 1991. *The origins of Modern Mind*. Cambridge: Harvard UP.

Merlin, D. 2001. *A mind so rare: The Evolution of Human Consciousness*. New York: Norton.

Norman, D. 1993. *Things That Make Us Smart*. New York: Addison-Wesley.

Quartz, S. R., Sejnowski, T. S. 2002. *Liars, Lovers and Heroes, What the New BrainScience Reveals About How We Become Who We Are*. New York: Harper and Collins.

Quéau, P. 2000. *La planète des esprits. Pour une politique du cyberespace*. Paris: Odile.

Salomon, G. 1993. *Distributed Cognitions: Psychological and Educational Considerations*. Cambridge: Cambridge UP.

Tomasello, M. 1999. *The Cultural Origins of Human Cognition*. Cambridge: Harvard UP.

4: WSIS and Organized Networks as New Civil Society Movements

NED ROSSITER

Introduction

In many respects, the material conditions of developing states have enabled the possibility of a range of conditions and experiences in advanced economies that could be considered as privileges constituted by legitimately enacted violence. Mary Kaldor notes that war and violence are both primary conditions for sustaining a civil society (see Kaldor, 2003: 31–38). As she writes: '*What Norbert Elias called the 'civilising process' – the removal of violence from everyday life within the boundaries of the state – was based on the establishment of monopolies of violence and taxation.*' (2003: 32) A monopoly of violence concentrates '*the means of violence in the hands of the state in order to remove violence from domestic relations.*' (Kaldor, 2003: 31–32) '*Modern sovereignty*', write Michael Hardt and Antonio Negri, '*was thus meant to ban war from the internal, civil terrain.*' (2004: 6)

The capture of violence by the state enables civil society to develop its key values of trust, civility, individual autonomy, and so forth, though within the framework of the rule of law as it is administered by the state. Moreover, the state's monopoly of violence minimises, though never completely eliminates, politically subversive elements and the possibility of civil war arising from within the territory of the nation. At a global level, the perversity of hegemonic states possessing a monopoly of violence operates as the basis upon which territorial sovereignty is maintained by way of subjecting violence upon alien states and their populations. A large part of this experience can be accounted for by referring to the histories of colonialism – a project whereby imperial states are able to secure the material resources and imaginary dimensions necessary for their own consolidation and prosperity.

Combining Hegel's thesis on the passage of nature/civil society/state with Foucault's notion of governmental power (i.e. the biopolitical, interpenetrative 'conduct of conduct'), political philosopher and literary theorist Michael Hardt defines civil society in its modern incarnation in terms of its capacity to organise abstract labour through the governmental techniques of education, training and discipline:

> *Civil society ... is central to a form of rule, or government, as Foucault says, that focuses, on the one hand, on the identity of the citizen and the process of civilization and, on the other hand, on the organization of abstract labour. These processes are variously conceived as education, training, or discipline, but what remains common is the*

active engagement with social forces (through either mediation or production) to order social identities within the context of institutions. (Hardt, 1995: 40)

With the governmentalization of the field of the social, a special relationship between civil society and the state is effected, one in which distinctions between institutions of the state and those of civil society are indiscernible, and where intersections and connections are diagrammatic. What, however, has happened to this constitutive relationship within our current era, one in which these sort of relationships have undergone a crisis as a result of new socio-economic forces that go by the name of neoliberalism? What sort of new institutions are best suited to the organization of social relations and creative labour within an informational paradigm? And what bearing, if any, do they have on inter-state and supranational regimes of governance and control?

In short, how do civil society movements articulate their values and how do they procure a multi-scalar legitimacy once the constitutive relationship between civil society and the state has shifted as the nation-state transmogrifies into a corporate state (or, in the case of developing countries, a state that is subject, for instance, to the structural adjustment conditions set by entities such as the World Bank and WTO)? Clearly, civil society values have not disappeared; nonetheless, the traditional modern constitutive framework has changed. Increasingly, civil society values are immanent to the socio-technical movements of networks. Issues of governance, I would suggest, are thus best addressed by paying attention to the technics of communication.[1] In the case of the WSIS project, this means shifting the debate from the 'multi-stakeholder approach' – which takes bureaucratically organized institutions (or networked organizations) as its point of departure – to one which places greater attention to the conditions of tension and dissonance as they figure with 'the political' of informationality. In other words, a focus on the materialities of networks and the ways in which they operate as self-organising systems would reveal quite different articulations that, in my view, more accurately reflect the composition of sociality within an information society.

Within a neoliberal paradigm we have witnessed what Hardt and Negri (2000) term *'a withering of civil society'* in which the structures and institutions that played the role of mediation between capital and the state have been progressively undermined. This shift has been enabled by the logic of deregulation and privatization, which has seen, in some respects, the socio-political power of both state and non-state institutions decline.[2] These include institutions such as the university, health care, unions and an independent mainstream media. For Hardt and Negri, the possibility of liberal democracy is seriously challenged by the hegemony of neoliberalism – or what they prefer to call the imperial, biopolitical and supranational power of 'Empire'[3] – since it threatens if not entirely eradicates traditional institutions of representation and mediation between citizens and the state. As Hardt and Negri write in their book *Empire*:

> *This withering can be grasped clearly in terms of the decline of the dialectic between the capitalist state and labour, that is, in the decline of the effectiveness and role of unions, the decline of collective bargaining with labour, and the decline of the representation of labour in the constitution. The withering of civil society might also be rec-*

ognized as concomitant with the passage from disciplinary society to the society of control. (2000: 328–329)

The society of control is accompanied by techniques of data-surveillance such as cookies, authcate passwords, data mining of individuals and their informational traces, CCTVs that monitor the movement of bodies in public and private spaces, and so forth. Some of these are related to the governance of intellectual property. New information and communication technologies (ICTs) thus play a key role in maintaining a control society. In an age of network societies and informational economies, civil society, or rather civil *societies*, have not so much disappeared as become reconfigured within this new socio-technical terrain in order to address problems immanent to the social, political and economic situation of mediatized life. Civil society, as it is resides within an informational plane of abstraction, continues to act as a key counter-force to and mediator between the state and capital. Thus, civil society does not entirely disappear or become destroyed with the onset of neoliberalism from around the 1970s–80s. Rather, there has been maintenance of civil society within our current network societies precisely because there has been a social desire and need to do so.

The emergent civil society movements go beyond satisfying the self-interest of individuals, as represented by consumer lobby groups, for example. Instead, they derive their affective and political power from a combination of formal and informal networks of relations. Think, for instance, of the effect the no-border refugee advocacy groups have had as observers of human rights violations meted out by the state. Whether one is for or against the incorporation of 'illegal immigrants' into the nation-state is secondary to the fact that civil society coalitions of activists, religious organizations and social justice advocates have played a primary role in constituting what Raymond Williams (1977) termed an emergent 'structure of feeling', or what can be thought of as the socio-technical organization of affect, that counters the cynical opportunism of populist conservative governments.

In an in-depth report entitled *Appropriating the Internet for Social Change*, Mark Surman and Katherine Reilly (2003a) examine the strategic ways in which civil society movements are using networked technologies. They identify four major online activities: collaboration, publishing, mobilization and observation. These activities are mapped along two axes: formal vs. informal and distributed vs. centralized (figure 1). Collaborative filtering and collaborative publishing, for instance, fall within the formal/distributed quadrant. Open publishing, mailing lists, research networks and collective blogs are located within the distributed/informal quadrant; personal blogs within the centralized/informal quadrant; and organizational website development, online petitions, online fundraising, e-membership databases and e-newsletters fall within the formal/centralized quadrant (Surman and Reilly, 2003b: 3). Surman and Reilly consider the '*tools that fall in the formal/centralized quadrant to be used primarily by large NGOs, unions and political parties.*' (2003b: 3) The logic of organization, production and distribution is, according to Surman and Reilly (2003b: 3), '*based on a "broadcast" model*' of communication. The distributed/informal quadrant, on the other hand, is more typical of activities undertaken by '*informal social movements, research networks and "virtual organizations*''. (Surman and Reilly,

2003b: 3) In this chapter, I will argue that it is time for 'informal social movements' and 'virtual organizations' – or what I prefer to call 'organized networks'-to make a strategic turn and begin to scale up their operations in ways that would situate them within the formal/centralized quadrant, but in such a manner that retains their informal, distributed and tactical capacities (see also Rossiter, 2004).

Figure 1: Strategic uses spectrum

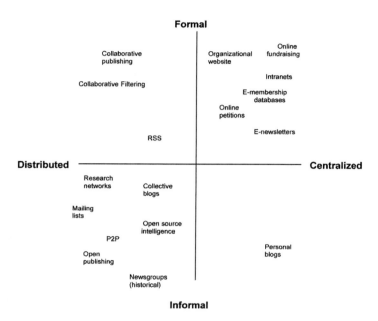

Source: Surman and Reilly (2003b: 3).

This chapter assesses the recent World Summit on the Information Society (WSIS) held in Geneva last December.[4] With disputes amongst various representatives over issues such as domain names, root servers, IP addresses, spectrum allocation, software licensing and intellectual property rights, the summit demonstrated that the architecture of information is a hugely contested area. As evidenced in official WSIS documents, consensus between governments, civil society groups, NGOs and corporations over these issues is impossible. Representation at the summit itself was a problem for many civil society groups and NGOs. As a UN initiative geared toward addressing the need for access to ICTs, particularly for developing countries, the problem of basic infrastructure needs such as adequate electricity supply, education and equipment requirements were not sufficiently addressed. Funding, of course, is another key issue and topic of disagreement.

Against this background, this chapter argues that the question of scale is a central condition to the obtainment and redefinition of democracy. Moreover, what models of democracy are global entities such as the WSIS aspiring to when they formulate future directions for informational policy? Given the crisis of legitimacy of rational consensus and deliberative models of democracy, this chapter argues that democracy within information societies needs to be rethought in terms of organized networks of communication that condition the possibility of new institutions that are attentive to problems of scale. Such a view does not preclude informational networks that operate across a range of scales, from sub-national to intra-regional to supra-national; rather, it suggests that new institutional forms that can organise socio-technical relations in ways that address specific needs, desires and interests are a key to obtaining informational democracy.

The 'multi-stakeholder' approach, as adopted in the WSIS process, in and of itself cannot fulfil the objective of, for example, *'an inclusive Information Society'*, as proposed in the official *Plan of Action* (WSIS, 2003). Despite the various problems associated with the WSIS, my argument is that it presents an important strategic opportunity for civil society movements: the 'denationalized' political legitimacy obtained at WSIS can, I would suggest, be deployed to political and economic advantage in the process of re-nationalization or re-localization. The emergence of organized networks as new institutional forms are best suited to the process of advancing the ambitions of WSIS.

Global Governance and the World Summit on the Information Society

The WSIS's two-stage meetings in Geneva, 2003, and Tunisia, 2005, exemplify the ways in which the political, social, economic and cultural dimensions of information and communication technologies afford civil society movements a political legitimacy in developments associated with issues of global governance that has hitherto been exclusive to supranational actors and multilateral institutions such as the WTO, the World Bank, IMF, the G8 nations, the UN, the OECD, APEC, ASEAN, NAFTA, and so forth. As the 'information society' has extended beyond the reserve of rich nations or advanced economies, actors such as the World Intellectual Property Organization (WIPO) and the Internet Corporation for Assigned Names and Numbers (ICANN) have emerged as institutions responsible for establishing common standards or information architectures that enable information to flow in relatively smooth, ordered and stable ways. Such entities have often been charged as benign advocates of neoliberal interests, as represented by powerful nation-states and corporations. As a UN initiative organized by the International Telecommunications Union (ITU), the WSIS has also been perceived by many as a further extension of neoliberal agendas into the realm of civil society. As Sasha Costanza-Chock reported in May 2003:

The ITU has always served governments and the powerful telecom conglomerates. Originally set up in 1865 to regulate telegraph standards, later radio, and then satellite orbit allocation, the ITU took on the Summit because it has recently been losing power to the telecoms that increasingly set their own rules and to the Internet Corporation for Assigned Names and Numbers (ICANN), which was created by the US government to regulate the Internet domain name system. The ITU is now facing heavy budget cuts and is desperate to remain a player in the global regulation of Information and Communication Technologies (ICTs). (Costanza-Chock, 2003)

The neoliberal disposition of ITU is further evidenced by the primacy given at the WSIS to issues such as cybercrime, security and electronic surveillance, taxation, IP protection, digital piracy and privacy (Yoshio Utsumi, Secretary-General of the ITU, cited in Constanza-Chock, 2003). The ITU's support of a summit concerned with bringing civil society movements into the decision making process of global information governance is one that is preconditioned by the empty centre of neoliberalism, which has seen governments in advanced economies reincorporating civil society actors and social organizations into matters of social welfare in the form of 'service providers'. Within a neoliberal framework, the interest of government and the business sector in civil society is underpinned by the appeal of civil society as a source of unregulated labour-power. This new axis of articulating civil society organizations through the logic of service provision functions to conflate 'civil society' with the 'private sector'. Such a conflation blurs or obscures what had previously been clear demarcations at the level of subjectivities, value systems and institutional practices. The conflation of civil society and the private sector is evident in much of the government documentation from the WSIS. In some ways this points to the multi-dimensional aspects of civil society – no longer can civil society be assumed to reside outside of market relations, for instance. In other ways, it raises the question of legitimacy: can civil society be 'trusted' when its condition of existence overlaps with market interests and needs of the private sector? Similarly, can the private sector be embraced by 'the Left' when the former displays credentials as a 'corporate-friendly citizen'? Indeed, what might 'citizenship' mean within a global framework? And then there is the mutually enhancing or legitimising function that such a convergence of actors produces: both civil society organizations and the private sector expand the discursive platforms upon which they stake out their respective claims. Ambiguities such as these point to the increasing complexity of relations between institutions, politics, the economy and sociality.

There is an urgent need to think through these issues and enact practices that go beyond the cynicism of Third Way style approaches to politics. The Third Way, as adopted by Blair, Clinton, Schroeder and others, is nothing but the expansion of market forces into social and cultural domains that hitherto held a degree of autonomy in terms of their articulation of different regimes of value (see Mouffe, 2000: 134–135; Scanlon, 2000). Moreover, there is a need for a radical pragmatism that engages civil society movements with economic possibilities in such a way that maintains a plurality of political ideologies, from Left to Right; this is something Third Way politics has undermined, the result being extremist manifestations of populist fundamentalism on both the Left and Right, but without the political institutions or processes to articulate their interests. The proliferation of terror is, in part, a symptom of this collapse in politics, a collapse which refuses the antagonisms that underpin the field of 'the political', and thus results in a situation whereby actors that might otherwise be adversaries instead become enemies. Since the antagonisms prevailing within information societies tend to be seen as distractions or debilitating to the WSIS project, I have doubts about the extent to which the 'multi-stakeholder' approach goes beyond some of the tenets of Third Way politics. Let us remember that communication systems are conditioned by the dissonance of information, or what Gregory Bateson (1972) termed *'the difference that makes a difference.'*

In an optimistic light, the 'multi-stakeholder' approach adopted at WSIS is indicative of a period of transition within supranational institutions. Yet paradoxically, the efforts of the ITU/UN to include civil society movements in the decision making process surrounding global governance of the information society is evidence of the increasing ineffectiveness of supranational governing and policy development bodies. The United State's cynicism and self-interest in bypassing the authority of the UN during the Iraq war and the breakdown of WTO summits in a post-Seattle climate of 'anti-globalization' protests are two extremes that point to the waning effectiveness of supranational institutions to address governance issues through international mechanisms. The expanding division and inequality in living and working conditions between the global North and the global South after successive WTO meetings and rounds of international agreements on trade liberalization is further evidence of the incapacity of supranational institutions to address the complexities of global governance.

In the case of the WSIS, Costanza-Cook (2003) maintains that the ITU's decision to organise the summit was partly motivated by their fear of redundancy as a governing body within an information society. Such a view is reiterated by Steve Cisler (2003) in his account of the tensions between ICANN and ITU at the WSIS:

ITU members like France Telecom and Deutsche Telekom long resisted the Internet. They were pushing Minitel, ISDN. African members saw (rightfully) how disruptive the Internet could be and resisted it.

The ITU was shocked by the growth of the Internet, and they have belatedly wanted to 'control' it. The failed WSIS proposal [to shift Internet governance away from WIPO] is just the latest attempt. Of course during this growing awareness of the importance of the Internet, the composition of the ITU has changed from almost exclusively government telcos (or PTT's) to a mix of old style government monopolies, dual government-private, and straight corporate telephone companies.

ICANN is a US government authorized non-profit corporation that is responsible for managing various technical aspects associated with Internet governance. These include *'Internet Protocol (IP) address space allocation, protocol identifier assignment, generic (gTLD) and country code (ccTLD) Top-Level Domain name system management [.com., .net, .org, etc.], and root server system management functions.'* (www.icann.org/general) The role of ICANN in Internet governance was disputed at the WSIS for a range of reasons. In his informative report commissioned by one of the more dominant civil society lobby groups at the WSIS, the Association for Progressive Communications (APC), Adam Peake (2004) unravels the debates that took place throughout the WSIS process about the role of ICANN in relation to issue of Internet governance. Peake notes that many were concerned that of the 13 root servers around the world that install all 'top level' domain name system (DNS) servers, 10 of these are located in the US (see Peak, 2004: 9). It becomes clear that at the level of technical infrastructure, the vertical stratification of the Net is shaped by geo-political, economic and cultural interests.

This tendency towards the vertical organization of information and its protocols rubs against the grain of efforts by civil society and open source movements to 'democratise' information and enhance the horizontalization of information management (see Galloway, 2004).

Alternative systems such as Anycast, which enable root servers *'to be 'cloned' in multiple locations'*, were proposed and implemented throughout the 2003 planning process of WSIS (Peake, 2004: 10). In other words, regional as distinct from US concentrated root servers are possible and came into effect in early 2003, but these only mirror or copy the US root servers and thus are not autonomously controlled.[5] Nevertheless, such alternatives begin to alleviate the concern that various civil society and government stakeholders had with respect to a perception that ICANN operates in the interests of maintaining a US control of the Internet, or at least supports the bias toward US Internet usage as represented by the location of root servers whose close geographic proximity to US-based users supports rapid response times on the Net.[6]

More significant concerns were raised about the gatekeeping role played by ICANN and the US government over the allocation of a country's top-level domain names. This was seen as undermining national sovereign control over domain names. Moreover, there is serious concern that the US Department of Commerce can potentially *'remove a country from the root, and therefore remove it from the Internet'*. (Peak, 2004: 10) It doesn't take much to imagine the devastating effect of removing a ccTLD in times of military and information warfare: a country's entire digital communications system is rendered useless in such an event, and social and economic impacts would come into rapid effect. A more likely scenario would see the US government intervening in the allocation of ccTLD's in instances of political or economic dispute. In this regard, the control of top-level domain names operates as a potential form of economic sanction or a real technique of unilinear leverage in business and political negotiations.

The other significant player in relation to this discussion of the WSIS is WIPO and their role in the global governance of information flows. WIPO is a UN agency with a mandate to 'harmonise' intellectual property rights across member states. In 1995 WIPO made an agreement with the World Trade Organization (WTO) to assist in facilitating the implementation of the TRIPS agreement across member states. More recently, WIPO's harmonization of patent law has been criticized for the way it restricts the degree of flexibility for and imposes substantial financial burdens on developing countries (Correa, 2004: 9). In a recent report written by Carlos Correa for the South Centre inter-governmental organization for developing countries, the following risks and asymmetrical aspects of WIPO's Patent Agenda for developing countries were summarized as follows:

> *... harmonized standards would leave little room for developing countries to adapt their patent laws to local conditions and needs; harmonization would take place at the highest level of protection (based on standards currently applied by developed countries, especially the United States and Western European countries) meaning that the process will exert an upward force on national laws and policies in developing countries resulting in stronger and more expansive rights of the patent holders with the corresponding narrowing of limitations and exceptions. Such higher standards are*

unlikely to have a positive effect on local innovation in developing countries; and also the danger that the current draft contains standards that are primarily aimed at benefiting the 'international industries' and not individual inventors or small and medium size enterprises. (Correa, 2004: 9)

Since it holds no legal authority at the national level, critics have frequently cast WIPO as an ineffective institution, although this is always going to be the case for a supranational institution whose legitimacy is as strong as the responsiveness to IPRs by member states. In instances where intellectual property protection is violated within national industries, as in the case of ongoing digital piracy of film and software within countries such as China, the lack of legal authority by WIPO is potentially offset by the mechanism of economic sanctions that can be imposed by adjacent supranational institutions and multilateral entities. A more substantial criticism of WIPO concerns its largely negative response to the issue of open source software and collaborative information flows that are best suited to developing countries without the financial resources to adopt proprietary informational systems. Thus the relationship WIPO holds with civil society movements and advocates of Open Source software and 'Open-Development' is often underpinned by conflicts in interest. Furthermore, the relationship between WIPO and the WTO casts the UN in the questionable role of advocating corporate interests over those of civil society.

This very brief overview of WIPO and some of the key issues associated with information architectures and the complex structural and institutional relationships begins to raise the question of what the relationship between global citizenship and Internet governance might mean within information societies. With stakeholders from civil society organizations, government and the private sectors, WSIS was never going to succeed as a global forum that seeks to be inclusive of diversity and difference if it was just going to focus on technical issues associated with Internet governance. The expansion of the debate on Internet governance, ICTs and issues of access and technical infrastructure to include civil society issues such as sustainability, funding, education, health, labour conditions and human rights functioned to sideline any centrality that ICANN and WIPO may have sought to hold during the summit. Many of the UN principles on human rights, for example, migrated into the *Civil Society Declaration* (2003) that came out of the summit. But like the *Universal Declaration of Human Rights* (1948), it is only as strong as the resolve of member states to ratify and uphold such principles within national legislative and legal frameworks.

The complexities of the WSIS process exceed the possibility of engaging their diversity. While the rhetoric has been one of inclusiveness, the experience for many working within civil society movements and, lest this chapter sound totally biased, the private sector, has been a frustrating one. As Adam Peake writes during one of the Prepcom meetings leading up the December summit in Geneva:

For those who don't know how WSIS works – everything happens at very short notice, situations have to be reacted to immediately, and it is very difficult for civil society to respond with the transparency and inclusiveness that we would hope. There simply is never time. [...]

I have one major concern. We should be very careful about how we raise issues around Internet governance in the WSIS process. We (civil society, private sector, Internet users) have a very weak voice in the process. WSIS is run by the States. Our only opportunity to speak, with *no* guarantee of being listened to, is in 1 or at best 2 10 minute sessions each day. ITU are the secretariat of the process and so have a very direct role in drafting text and framing arguments for the States to consider [...] (Peake cited in Byfield, 2003)

Critical Internet researchers have also had cautious words to say about the extent to which the civil society activists – or what many now refer to as the 'multitudes', or movement of movements – can expect to make a substantial impact on the WSIS process. Again, the diversity of stakeholders and their competing interests brings in to question the ambitions of the 'multi-stakeholder' approach. If dissonance is taken as the condition of informationality, as distinct from deliberation and consensus as idealistic outcomes, then we begin to orient ourselves around the possibility of 'post-representative' systems of organising socio-technical relations. Co-moderator of the nettime mailing list, Ted Byfield, gives his perspective in a posting in March 2003 – around the time the ITU began to soften its tone of market-oriented, technical-driven solutions to Internet governance:

My own view is that the activists who think the ITU/WSIS process is just another three or four-letter target for generic social-justice demands should be much more sensitive to the context [....]. the logic of 'multitudes' may not be representative, but the logic of monolithic organizations (at best) *is* representative, so it would be a mistake to assume that the delirious logic of the movement of movements will somehow transform the ITU into some groovy, polyvocal provisionalism. it won't. One of the most 'progressive' things the WSIS process can accomplish is to minimize the scope of ITU activities. Cookie-cutter activist demands will inevitably put pressure on the ITU to expand its purview – and provide a pseudo-legitimating cover for such expansions. This would NOT be a good thing. (Byfield, 2003)

As it turned out, the *Declaration of Principles* (WSIS, 2003) and *Plan of Action* (WSIS, 2003) articulate exactly what Byfield fears: lip service to concerns of civil society movements, which are beyond the scope of the bureaucratically driven governance structures of nation-states, who are incapable of dealing with complex social and cultural issues. This situation will inevitably result in a Tunis 2005 summit that skirts around the serial incapacity of participating governments to implement many, if any, of the recommendations proposed in the *Plan of Action*. Perhaps the best thing the WSIS could do is stick with a relatively limited agenda. That might mean keeping the debate on the information society focussed on limited technical and legal issues-policy domains that nation-states do have some capacity to at least administer. It would, however, be a disaster to see Internet governance shifted in any exclusive way into the regulatory domain of nation-states. It is unclear at a technical level how necessary it is for a supranational, global institution to be steering Internet governance issues. The question that has persisted throughout WSIS is whether ICANN is the body best suited to this task. Currently, ICANN appears to have short time left to live.

It was inevitable that the broad, inclusive ambitions of the WSIS at the end of the day turned into a rhetorical machine. While this has meant that civil society movements have obtained a degree of legitimacy at a supranational, institutional level, it is highly doubtful whether the WSIS itself is able to turn the tables on the broad and complex social situations that inter-relate with ICTs. The legitimacy obtained by civil society movements involved in the WSIS process can be transferred as political and symbolic leverage within other, more focussed platforms at national and translocal levels. This process of a re-nationalization of the discursive legitimacy of civil society concerns and values is the next challenge.

All of this background summary, minimal and reductive as I have presented it, finally brings me to the crux of my argument in this chapter: democracy within an informational society is challenged, perhaps more than anything, by the problematic of scale and the ways in which cumbersome, top-heavy and bureaucratic-driven supranational institutions involved in issues of global governance are always going to fail. From the WSIS emerges a pattern indicating that governing institutions have substantial limits in terms of policy development that acts as a driver of democratic change. Such a problematic is one of scale. It also has much to do with the correspondence between institutional temporalities and the limits of practice. The temporal rhythms of the networked organization, as distinct from organized networks, are simply not well suited to the complexities of socio-technical relations as they manifest within informational societies. Despite the impact of post-Fordist techniques of re-organising institutional relations and modes of production, the networked organization persists as the dominant environment within which sociality is arranged. Such institutional formations will only continue to struggle to keep apace with the speed of transformation and the contingencies of uncertainty peculiar to the informatization of social relations.

Institutional Scale and the Technics of Governance

At best, the 'informational citizen' is one who has recourse to representative systems of governance adopted by liberal democratic nation-states. But it is well and truly time to invent new post-political, non-representative models of democracy. The crisis of liberal democracy across the West over the last twenty to thirty years is carried over to the debates occasioned by the WSIS. The distributive, non-linear capacity of the Net shapes social-technical relations and information and knowledge economies in ways that do not correspond with the old, hierarchical structures and governance processes peculiar to the modern era. The challenge of organization and governance is intrinsically bound to the informatization of the social. Representative models of democracy do not correspond with this situation.

While it may appear as just an institution whose exclusive responsibility concerns technical architectures of Internet governance, the case of ICANN points to more substantial matters associated with models of global governance within an age of networks. Described by some as *an experiment in democratic governance on a global scale* (Palfrey, 2004: 411–412), ICANN embodies many of the challenges facing organized networks, both in terms of how they understand themselves and how they function. The contest over ICANN's monopoly of Internet governance – as raised by civil society concerns at WSIS, the interests of the ITU as a new player in Net governance, and the ambitions of the EU as a 'second-tier' super-state-signals not just the difficulties associated with 'multi-stakeholder' approaches to governance; more than anything, the ICANN story points to the profound mistake in assuming the Net can reproduce the pillars of 'democracy' in its idealized *'Westphalian international order'*. (Bensaid, 2003: 317) John Palfrey (2004: 412) charts the history of ICANN and what he sees as its imminent demise. ICANN, he writes, *'sought to empower the Internet user community, including the private sector, to manage a system necessary for the stable operation of the Internet.'* So far so good. Things became unstuck, however, at a structural level in terms of incorporating a range of stakeholders into the decision-making process of ICANN:

> *Its novel, though ultimately flawed, structure has enabled a coalition of private-sector interest groups to manage the domain name system ('DNS') with broad input from individual users and limited but growing input from nation states. However, ICANN has failed to attract and incorporate sufficient public involvement to serve as the blueprint for building legitimacy through the Internet. Those who sought through ICANN to prove a point about democracy have misplaced their emphasis, because ICANN's narrow technical mandate has not lent itself to broad-based public involvement in the decision-making process.* (Palfrey, 2004: 412)

And:

> *ICANN has sought to legitimate itself as an open and representative body, striving toward a bottom-up decision-making processes grounded in consensus and inclusion.* (Palfrey, 2004: 412–413)

The online global election in 2000 of five 'At Large' members of the 19 member directorship is a great example of the mistaken understanding of what constitutes a representative polity within a global information society that is defined, from the outset, by an uneven geography of information. Who, for instance, are the elected five members (to say nothing of the 14 unelected members) supposed to represent? Their nation-state of origin? A particular set of issues? And who is 'the public' that participates in such events?[7] These are all questions that lead to one conclusion: attempts at reproducing a modern socio-technics of representative democracy within an informational plane of abstraction can only result in failure. The valorization of 'openness' is not a particularly helpful libertarian mantra to maintain when dealing with the uneven geography of information.[8]

In case we've forgotten, such speculative discourses are ones associated with the 'New Economy', and we saw what that amounted to when the dotcom bubble burst and NASDAQ high-tech stocks crashed in April 2000: a spectacular tech-wreck that resulted in pretty much instant bankruptcy and overnight unemployment for many (see Frank, 2000; Henwood, 2003; Lovink, 2003: 56–85). The religious faith that IT development is synonymous with instant and sustainable growth was certainly brought into question with the massive devaluation of dotcoms and telcos. But one could be forgiven for wondering if the monumental tech-wreck ever happened. Government and education institutions have been particularly slow to awaken to the fact of the NASDAQ collapse. The rhetoric of 'e-solutions' as the answer to all problems continues to run thick in these places. Part of the reason for this has to do with the way in which the deregulation of many government and education institutions has followed on from the deregulation and privatization of telcos and the media industries, which was fuelled by the market hype of what critical Internet theorist Geert Lovink (2003) calls *'dotcom mania'*. In other words, the ongoing hype generated out of the IT sector seems to be the only discursive framework available for countries enmeshed in the neoliberal paradigm, be they advanced economies in the West or countries undergoing a 'leap-frogging' of modernity (see Rossiter, 2002b). While the WSIS forums have been successful in generating a new legitimacy for civil society values, too often one is reminded of the deeply unimaginative ideas driving the ambition for an inclusive information society.

Jeanette Hofmann – one of the elected At Large members of ICANN – recounts a key problem confronting organizations as they scale up their level of operations. Speaking of the paradox that comes with obtaining legitimacy within international institutions, Hofmann observes: *'As soon as civil society organizations assume formal roles in international forums, their representativeness and legitimacy are also called into question. Ironically, NGOs are charged with the democratic deficit they once set out to elevate.'* (Lovink and Hofmann, 2004; see also Rossiter, 2002b) This notion of a 'democratic deficit' can be extended to the Association for Progressive Communications (APC/www.apc.org), who have been one of the peak lobby groups within the Internet governance and communication rights debates associated with WSIS. The effect of an increased institutional and discursive visibility is, of course, conditioned by an increased marginalization of other civil society actors. Again, this points to the limits and problem of politics that operates within a representative framework, which the APC presupposes as its mandate of governance. The APC story is also symptomatic of the structural logic of political pragmatism within a multi-stakeholder, trans-scalar supranational policy forum such as WSIS.

As I am arguing in this chapter, it is time to invent non- or post-representative modalities of organization, as distinct from representative idioms of governance (see also Rossiter, 2004; Virno, 2004). In this way, the technics of communication is granted the kind of primacy that corresponds with the informatization of sociality. Moreover, the disjuncture between, if you will, the signifier and signified (i.e. speaking positions) is sidelined in favour of collaborative and distributive technics of composition. Do not get me wrong; in no way am I proposing some kind of naïve 'ideal speech act' here. There should be no illusion that distributive networks are somehow free from vertical systems of organization, be they symbolic or material. Rather, the technics of communication within a digital era do not correspond with the kind of institutional arrangements that persist within debates on the 'information society' and presupposed in the 'multi-stakeholder approach' of WSIS. These kinds of institutions can be understood as networked organizations. They are clumsy when it comes to the management of information.

ICANN faced a similar difficulty to that of civil society organizations, as identified by Hofmann. But what I've been suggesting is that the problematic of 'democratic practice' goes beyond the level of discursive legitimacy. More fundamentally, there is a problem with the way in which principles of democracy peculiar to the modern state system are translated into the socio-technical environment of the Internet. The result is always going to be failure. Completely new understandings of organizational structures, practices, and political concepts are called for with the emergence of organized networks in order to create value systems and platforms of legitimacy that are internal to networks. As I briefly sketch in my concluding comments to this chapter, the concept of a 'processual democracy' offers one possibility for exploring alternative political formations that are attentive to the ways in which practice is situated within the media of communication.

The case of ICANN serves as a parallel instantiation of the kind of governance problematics faced throughout the stages of WSIS. The WSIS process embodied a shift in relations between the UN and non-state actors, which, for the past decade or so had been characterized by a 'top-down' approach by which the UN engaged NGOs (see Padovani & Tuzzi, 2004). In their recent report on the WSIS for the Social Science Research Council, New York, Claudia Padovani and Arjuna Tuzzi (2004) consider such a mode of governance as 'institutional'. By contrast, they see the 'bottom-up' or 'globalization from below' approach at the WSIS as a challenge to earlier relations between the UN and civil society actors. Both, they argue, were operating during the WSIS and the two-year lead up of preparatory committee meetings (PrepComs), regional conferences and follow-up meetings.

At a reductive level, the differences between these two approaches are apparent in the range of documentation and critical responses to come out of the summit. The two approaches are most clearly delineated in their articulation of values and modes or processes of governance. In terms of values, the institutional approach embodied by government and business representatives was predominantly interested in market-based and technically oriented solutions to ICTs and their relationship to issues of global governance. In effect, government and business participants reproduce the neoliberal paradigm that has dominated the past two decades of government policy-making in the West. Here, one finds the international lingua franca of policy that adopts an instrumentalist faith and technologically determinist simplicity to the uneven and situated problems of social, cultural and economic development.

For example, in the government *Plan of Action* there is an emphasis on technical infrastructures and informational access functioning as the primary enabling devices for 'universal education' and 'lifelong learning'. This sort of Third Way rhetoric is further compounded in the *Plan of Action*'s discourse on 'capacity building' – a phrase shared amongst a range of WSIS stakeholders and common to many civil society organizations, but one that is understood in terms of 'e-learning' and 'distance education' in the *Plan of Action*. Such phrases are firmly entrenched within neoliberal discourses that understand education as a unilinear, hypodermic communication process driven by service providers operating under the auspices of imperialist political economies. Within a dotcom paradigm, such discourses amount to no more than boosterism for the IT sector (see Lovink, 2003: 57–85). The economic and political pressures faced by the university sector in the West contribute to a dependency relationship within indigenous education systems in developing countries. 'E-learning' and 'distance education' are heavily promoted as the financial panacea for cash-strapped universities in the West, and the 'consumer' of such projects frequently consists of countries without nationally developed educational infrastructures. The need by developing countries for external providers of education is then often used as the justification for developing IT infrastructures. Education becomes subject in the first instance to the interests of market economies, and policy developments associated with civic values are then articulated in economistic terms. Throughout the *Plan of Action*, policy initiatives are driven by the capacity for governments to index access against targets and performance indicators. Such a technique of governance and decision-making is symptomatic of the limits of supranational entities to deal with complexity and functions to give the false impression of 'demonstrable outcomes'.

The 'bottom-up' approach, as represented by civil society organizations, NGOs and activists, was much more concerned with ensuring that social and cultural priorities were embraced in the *Declaration of Principles* and *Plan of Action*. Civil society movements have been effective in shifting the WSIS agenda from a neoliberal, technologically determinist set of proposals to a more broad understanding of an information society that is preconditioned by the materialities of communication. The 'multi-stakeholder' approach that emerged out of the WSIS meetings to date has enabled issues of concern to civil society movements to migrate into the field of supranational policy-making. The two primary documents produced so far are clear on one thing – a technological fix to social and economic problems is not going to work.

The reason such a 'discourse war' between top-down and bottom-up approaches to information governance was so significant is that the success of the WSIS process in ensuring a 'social justice and development' agenda for civil societies and their relationship with ICTs in many ways rests with governments adopting the principles and proposals outlined in the official documentation. Many oral and written submissions to the drafting of the official *Declaration of Principles* and *Plan of Action* were left out of the final documents. The decoupling of 'macro' and 'micro' actors was further reflected in the summit itself, with activists, grass-roots organizations and NGOs running meetings and workshops in parallel to the official UN program for the Geneva meeting (see Padovani and Tuzzi, 2004). Padovani and Tuzzi suggest a much more overlapping approach characterized the summit. Certainly, WSIS has presented its own peculiarities with regard to the problematics of process, decision-making and identification of key issues (see Betancourt, 2004). But one should not see WSIS as exceptional or unique in terms of organising a range of stakeholders around a particular theme or issue perceived as having international significance. The UN, after all, has a history of hosting approximately one summit per year since the 1992 Earth Summit (Conference on Environment and Development) in Rio de Janeiro (see Klein, 2003: 3).

It would thus be a mistake to see the 'multi-stakeholder' approach to governance at the supranational level as exceptional. Arguably, all summits have had to address the challenge of managing a range of stakeholders and their competing interests and situations. What distinguishes WSIS from previous summits is the ways in which the process of informatization has interpenetrated the organization of social relations, economic modes of production and systems of communication. Such a situation does indeed call for new models of governance, but whether the idea of 'multi-stakeholder governance' in and of itself is sufficient to the task of socio-technical complexity is, I would suggest, doubtful. A substantial challenge to this model consists of the highly variable dimensions of power and its operation across a range of scales and a diversity of actors. As Padovani and Tuzzi (2004) maintain, *'the 'multi-stakeholder approach' is not yet a model and needs to be defined, not only in theory but in practice, taking into consideration the nature and level of power the different stakeholders can exercise.'*

Conclusion

It is time to develop a model of democratic polities that engages, in the first instance, with the condition of immanence that is peculiar to socio-technical relations as they are arranged within information societies. Elsewhere I have advanced the concept of 'processual democracy' as one that corresponds with new institutional formations peculiar to organized networks that subsist within informationality (Rossiter, 2004). A processual democracy unleashes the unforeseen potential of affects as they resonate from the common of labour-power. A processual democracy goes beyond the state-civil society relation. That relation no longer exists, at least not in terms of its traditional bi-modal structure.

Processual democracies necessarily involve institutions, since institutions function to organise social relations. Processual democracies also continue to negotiate the ineradicability of antagonisms. Their difference lies in the affirmation of values that are internal to the formation of new socialities, new technics of relations. Certainly, they go beyond the limits of resistance and opposition – the primary activity of tactical media and the 'anti-corporatization' movements. This is not to dispense with tactics of resistance and opposition. Indeed, such activities have in many ways shaped the emergence of civil society values into the domain of supranational institutions and governance, as witnessed in the WSIS debates. A radical adaptation of the rules of the game is a helpful way of thinking the strategic dimension of processual democracies.

Organized networks are the socio-technical system best suited to further develop the possibility of an inclusive information society. Since they have the capacity to operate on multiple scales of practice and communication, the challenge for organized networks consists of how they will engage their counterpart-networked organizations – which, after all, are the dominant institutions. One of the first tasks for organized networks is to address the question of sustainability. Only then can they begin to provide an operative base for their subnational, intra-regional and transnational geographies of expression.

Notes

1   Andrew Murphie (2004: 136) defines the term 'technics' *'as a combination of technologies, systematic processes and techniques, whether these are found in the organization of living or non-living matter.'* I will adopt this sense of technics throughout this chapter. See also Mumford (1934), Latour (1993) and May (2002: 28–35).

2   Although the so-called 'decline' of state sovereignty and non-state institutions is peculiar to a modern era of sovereignty. I maintain that state sovereignty has transformed rather than disappeared. Similarly, the role of non-state institutions can be considered in terms of emergent civil society movements.

3   In his biographical and biopolitical abecedary undertaken in collaboration with Anne

Dufourmantelle, Negri defines 'Empire' even more precisely as: *'the transfer of sovereignty of nation-states to a higher entity.'* (Negri, 2004: 59)

4   For background information and critical reports on the WSIS, see http://www.itu.int/wsis, http://www.wsis-online.org, http://www.unicttaskforce.org, http://www.apc.org, http://www.ssrc.org, http://www.southcentre.org. See also reports and debates on nettime (http://www.nettime.org) and incommunicado (http://www.incommunicado.info). See also Betancourt (2004).

5   The kind of regionalism constituted by the cloning of root servers raises another interesting issue: namely, the geography of power that attends the complex multi-layered dimensions of competing 'regionalisms'. How, for example, does the informational regionalism of the Anycast system reproduce or contest more established regional formations of transnational cultural flows and the diaspora of labour-power, or the regionalisms of multi-lateral trade agreements and economic blocs, or the sub-national, intra-regional formations of civil society movements?

6   Peake notes that *'The request of the WSIS Plan of Action to deploy 'regional root servers' was achieved even before the Summit was held.'* (2004: 10) The question remains as to whether this plan is put into effect-something that will unfold in the lead-up to the 2005 Summit.

7   Similarly, as Antonio Negri has noted, *'... the problem is that the term 'democracy' has been emptied of all its meaning. Democracy is said to be identified with 'the people' – but what is the people?'* (2004: 117).

8   For example, many libertarians and activists insist that intellectual property (IP) laws should be universally abolished, since IP inscribes a regime of scarcity upon that which is digitally encoded and thus remains undiminished at the level of form when it is reproduced and distributed. Certainly, there are strong reasons to support such a position. There is a great need to combat the substantial financial and legal barriers that emerge with accessing information and knowledge resources associated with patents for agricultural development and vaccinations. However, there are many factors overlooked in any blanket approach to the problem of intellectual property. For an argument of how intellectual property regimes hold the potential to advance indigenous sovereignty movements in Australia, see Rossiter (2002a).

References

Bateson, Gregory. 1972. *Steps to an Ecology of Mind*. New York: Ballantine Books.

Bensaid, Daniel. 2003. 'Sovereignty, Nation, Empire', trans. Isabel Brenner, Kathryn Dykstra, Penny Oliver and Tracey Williams, pp. 317–323 in William F. Fisher and Thomas Ponniah (eds.) *Another World Is Possible: Popular Alternatives to Globalization at the World Social Forum*. London and New York: Zed Books.

Betancourt, Valeria. 2004. 'The World Summit on the Information Society (WSIS): Process and Issues Debated', Association for Progressive Communications (APC), http://www.apc.org.

Byfield, Ted. 2003. 'ccTLDs, WSIS, ITU, ICANN, ETC', posting to nettime mailing list, 7 March, http://www.nettime.org.

Cisler, Steve. 2003. 'ICANN or UN?', posting to nettime mailing list, 12 December, http://www.nettime.org.

WSIS Civil Society Plenary. 2003. *Shaping Information Societies for Human Needs. Civil Society Declaration to the World Summit on the Information Society*, Geneva, http://www.itu.int/wsis/docs/geneva/civil-society-declaration.pdf.

Correa, Carlos M. 2002. 'The WIPO Patent Agenda: The Risks for Developing Countries, Trade Related Agenda, Development and Equity (T.R.A.D.E.)', *Working Chapters*, no. 12, South Centre, http://www.southcentre.org

Costanza-Chock, Sasha. 2003. 'WSIS, the Neoliberal Agenda, and Counter-proposals from "Civil Society"', posting [by Geert Lovink] to nettime mailing list, 12 July, http://www.nettime.org.

Frank, Thomas. 2000. *One Market Under God: Extreme Capitalism, Market Populism, and the End of Economic Democracy*. New York: Anchor Books.

Galloway, Alexander. 2004. *Protocol: How Control Exists After Decentralization*. Cambridge, Mass.: MIT Press.

Hardt, Michael. 1995. 'The Withering of Civil Society', *Social Text* 14(4): 27–44.

Hardt, Michael, Negri, Antonio. 2000. *Empire*. Cambridge, Mass.: Harvard University Press.

Hardt, Michael, Negri, Antonio. 2004. *Multitude: War and Democracy in the Age of Empire*. New York: Penguin Press.

Henwood, Doug. 2003. *After the New Economy*. New York: The New Press.

ICANN, http://www.icann.org.

Kaldor, Mary. 2003. *Global Civil Society: An Answer to War*. Cambridge: Polity.

Klein, Hans. 2003. *Understanding WSIS: An Institutional Analysis of the UN World Summit on the Information Society*, Internet & Public Policy Project, School of Public Policy, Georgia Institute of Technology, http://IP3.gatech.edu.

Latour, Bruno. 1993. *We Have Never Been Modern*, trans. Catherine Porter. Cambridge, Mass.: Harvard University Press.

Lovink, Geert. 2003. *My First Recession: Critical Internet Culture in Transition*. Rotterdam: V2_/NAi Publishers.

Lovink, Geert, Hofmann, Jeanette. 2004. 'Open Ends: Civil Society and Internet Governance [Interview]', posting to nettime mailing list, 12 August, http://www.nettime.org.

May, Christopher. 2002. *The Information Society: A Sceptical View*. Cambridge: Polity.

Mouffe, Chantal. 2000. *The Democratic Paradox*. London: Verso.

Mumford, Lewis. 1934. *Technics and Civilization*. London: Routledge & Kegan Paul.

Murphie, Andrew. 2004. 'The World as Clock: The Network Society and Experimental Ecologies', *Topia: A Canadian Journal of Cultural Studies* 11(spring): 117–139.

Negri, Antonio with Dufourmantelle, Anne. 2004. *Negri on Negri*, trans. M. B. DeBevoise. London and New York: Routledge.

Padovani, Claudia, Tuzzi, Arjunna. 2004. *Global Civil Society and the World Summit on the Information Society: Reflections on Global Governance, Participation and the Changing Scope of Political Action*, Social Science Research Council, New York, http://www.sscr.org/programs/itic/publications/knowedge-report/memos/padovani3-4-30-04.pdf.

Palfrey, John. 2004. 'The End of the Experiment: How Icann's Foray into Global Internet Democracy Failed', *Harvard Journal of Law & Technology* 17: 409–473.

Peake, Adam. 2004. 'Internet Governance and the World Summit on the Information Society (WSIS)', Association for Progressive Communications, http://rights.apc.org/documents/governance.pdf.

Rossiter, N. 2002a. 'Modalities of Indigenous Sovereignty, Transformations of the Nation-State, and Intellectual Property Regimes', Borderlands E-Journal: *New Spaces in the Humanities* 1(2), http://www.borderlandsejournal.adelaide.edu.au/issues/vol1no2.html.

Rossiter, Ned. 2002b. 'Whose Democracy? Information Flows, NGOs and the Predicament of Developing States', Dark Markets: Infopolitics, Electronic Media and Democracy in Times of Crisis, International Conference by Public Netbase/t0, Muesumsplatz, Vienna, 3–4 October. http://darkmarkets.t0.or.at/.

Rossiter, Ned. 2004. 'Virtuosity, Processual Democracy and Organized Networks', The Italian Effect: Radical Thought, Biopolitics and Cultural Subversion, Sydney University, 9–11, September, 2004, http://www.arts.usyd.edu.au/departs/rihss/italianeffect.html.

Scanlon, Chris. 2000. 'The Network of Moral Sentiments: The Third Way and Community', *Arena Journal* 15: 57–79.

Surman, Mark, Reilly, Katherine. 2003a. *Appropriating the Internet for Social Change: Towards the Strategic Use of Networked Technologies by Transnational Civil Society Organizations*, version 1.0. New York: Social Science Research Council. Available at http://www.ssrc.org/programs/itic/.

Surman, Mark, Reilly, Katherine. 2003b. *Executive Summary. Appropriating the Internet for Social Change: Towards the Strategic Use of Networked Technologies by Transnational Civil Society Organizations*, version 1.0. New York: Social Science Research Council. Available at http://www.ssrc.org/programs/itic/.

Virno, Paolo. 2004. *A Grammar of the Multitude*, trans. James Cascaito Isabella Bertoletti, and Andrea Casson, forward by Sylvère Lotringer. New York: Semiotext(e). Also available at: http://www.generation-online.org/c/fcmultitude3.htm.

Williams, Raymond. 1977. 'Dominant, Residual and Emergent', pp. 121–127 in *Marxism and Literature*. Oxford: Oxford University Press.

World Summit on the Information Society (WSIS). 2003. *Declaration of Principles*, Geneva, 12 December, http://www.itu.int/dms_pub/itu-s/md/03/wsis/doc/S03-WSIS-DOC-0004!!PDF-E.pdf.

WSIS. 2003. *Plan of Action*, Geneva, 12 December, http://www.itu.int/dms_pub/itu-s/md/03/WSIS/doc/S03-WSIS-DOC-0005!!PDF.pdf.

5: How Civil Society Can Help Civil Society

STEFANO MARTELLI

Introduction

Today many different definitions have been offered in sociology about the nature of contemporary society: 'advanced modernity' (Giddens, 1990), 'reflexive modernity' (Beck, Giddens & Lasch, 1994), 'liquid modernity' (Bauman, 2000), 'post-modernity' (Lyotard, 1979; Jameson, 1984; Donati, 1997; Martelli, 1999), etc. Despite these differences, a trait is common to all definitions: the recognition of the increasing importance of media. As a result, contemporary society is the first 'mediated' society in history.

The process of mediatization of society brings about a lot of consequences: the globalization of informations, the transformation of politics and leadership, the digitalization of the industrial production, the need for media education and so on. It is too hard to establish whether this mediatization of society is an *advancement* of society on its way to progress – in the universalistic sense of the term, indeed. Changes are in progress *now*, and it is difficult to think beyond the flow in which we all swim.

Communication has acquired an increasing importance in all sectors of society: the state and its administration, the market and its enterprises, the 'third sector' and the many organizations, which operate in the space between the first two sectors. The 'third sector' is the space in which civil society organizes and produces itself; it is shaped by the network of private-social organizations (PSOs), which promote the life quality of a community and help the poor, the sick and the weak (Donati, 1993b; Donati & Colozzi, 2002; 2003).

Since the public sphere is today a 'mediated' sphere, the 'third sector' too has to communicate. In fact, communication is the pre-condition for the visibility of all organizations; but, while the state and the market have their own channels and are often given great attention by the media, the 'third sector' has few channels of its own and basically receives no attention by the media. As a result, it risks being invisible to everybody's eyes.

In this chapter, I will try to describe an experiment of communication through information and communication technologies (ICTs), which is still in progress in Palermo (I).[1] On November 2003 four civil society institutions founded the *Telematic Portal for the communication of the 'third sector' in Palermo* (the *Portal*), in order to promote the visibilization of the pro-social activities carried out by the PSOs.

First, I will try to describe the social nature and the composition of the 'third sector' in Italy. I will also point out the PSOs' need to communicate within the 'mediated' public sphere. Secondly, I will show the first results reached by a research action on the communication activities of the Sicilian PSOs carried out by a team of sociologists of the University of Palermo as a part of a national network of several Italian universities working on the 'third sector' and the social capital in Italy (Donati & Colozzi, 2004). Finally, I will describe the *Portal* and the role of the 'communication account', a new figure that may give some help to the 'third sector' in its efforts to communicate better.

Between the State and the Market: The Quest of Private-Social Organizations (PSOs) to Communicate within the 'Mediated' Public Sphere

What is There Between State and Market?
As many sociologists have pointed out, a 'third sector' exists between the state and the market. Therefore, the social actors of the welfare politics are

- the state (the *first* sector of society), which offers *public* goods and services;

- the market (the *second* sector), which offers *private* goods and services;

- the *third* sector, which is the most organized part of civil society – it offers a new type of goods and services, i.e. the *relational* ones (Donati, 1993b; Donati & Colozzi, 2002; 2003).

Relational goods and services are not *material*. They are produced by the pro-social action and enjoyed by both the members of PSOs and the *Alter* according to a vision of life based on reciprocity (gift, mutual help, wide exchange, etc.).

Indeed, the social relation *is* the main good – the most important resource for the people living in a post-modern society. Without the *Alter*, I cannot produce trust, well-being and a sense of belonging to any community. In order to produce relational goods, many people gather in the various groups and associations of the 'third sector' – i.e the PSOs.

What are the Pillars of the 'Third Sector'?
There are five pillars of civil society which shape the 'third sector' (see figure 1):

- Associations for the social promotion;
- Associations of families;
- Organizations of volunteers;
- Social cooperatives;
- Social foundations.

All the organizations belonging to these five pillars are PSOs. They produce those relational goods (offered neither by the state nor by the market), which are necessary for people's well-being and for an increasing quality of life in the whole society. The number of PSOs operating in each pillar as well as the type of their activity are quite different, as you can see from figure 1.

Figure 1: The five pillars of the 'third sector' in Palermo (I)

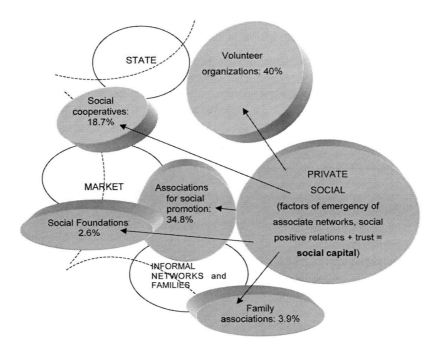

Source: my elaboration from Donati and Colozzi (eds. 2004: 16).

Figure 1 shows the position of the 'third sector' between the state, the market, and the informal networks and families operating in the private-social sphere; it also shows the size of the different pillars in Palermo. Three out of four PSOs are voluntary organizations (40%) or associations for the social promotion (34.8%); the remaining are social cooperatives (18.7%), family associations (3.9%) and social foundations (2.6%). This distribution is quite similar in the largest Italian cities.

'Third Sector' or No-Profit Organizations?
Is 'third sector' an equivalent word for no-profit organizations (NPOs)? The answer is negative: social foundations, a lot of family associations and some volunteer organizations are not NPOs, therefore the term 'third sector' stands for a more general concept than NPOs.

The ISTAT (Italian Institute of Statistics) has estimated more than 210,000 active PSOs (Istat, 2000), but the total number of 'third sector' organizations is much higher. In Italy about 1,000,000 volunteers are involved in pro-social activities, therefore the whole 'third sector' most probably amounts to about 3,500,000 people.

A Research Action on the Communication Activities of the PSOs in Palermo

A Recent Survey in Italy about the 'Third Sector'

In the years 2001–2002, six Italian universities[2] carried out a national survey on the culture and the organization of the 'third sector' in Italy. This research was approved and co-financed by the Miur, the Italian Ministry of Education University and Research. The national sample was formed by 2.326 PSOs' individual members and by 588 PSOs; it was the first statistically representative sample of the 'third sector' in Italy. For the final findings, see the book edited by Donati and Colozzi (2004).

One of the most important findings obtained by this survey shows that the PSOs would like to communicate, but they do not know how to do it. In fact a lot of them do realize that in the society of global communications this activity is necessary, yet they often have neither the tools nor the communicative competence to do it successfully (Martelli, 2004).

The research activity on the Italian 'third sector' is still going on. In the period 2003–2004 one more Italian university – the University of Padova (North Italy) – joined the first six ones. The common goal is to study the social capital in Italy and the local dynamics of the 'third sector'. This research program, too, as the previous one, has been approved and co-financed by the Miur. More specifically, the research program of the University of Palermo intends both to study the PSOs of this large town (about 700,000 inhabitants) and support their effort to communicate better through the information and communication technologies (ICTs).

Some Findings on the Communication Activities of PSOs in Palermo

In this first phase of our research action in Palermo, my research team and I contacted more than 150 PSOs. We gave to the managers of these PSOs three different questionnaires to fill out and to return to us.

i) a questionnaire to describe their values and attitudes, and the social goals of their voluntary action (individual data);

ii) a questionnaire to describe the organization of the PSO (structural data);

iii) a questionnaire to describe the communication activities of the PSO (organizational data).

Whereas the first two questionnaires are the same as those used in the national survey (2002), the third one is a new and original contribution of the University of Palermo to the national research on the social capital in Italy.

The Communication Activities of the PSOs in Palermo

Less than one out of ten PSOs puts *communication* at the top of its activity planning. As you can see from table 1, for the PSOs in Palermo the most important activities are: *internal organization* (30%), *formation and refresher courses*

(29.1%), *planning and development of new services* (18.5%). Only a few of them (8.1%) regard communication as an important activity. *Fund-raising, Relationships with public administrations* and *Recruitment of new members* are regarded as important activities by even less PSOs (4.7%) and so on.

Table 1: The importance of communication within the activities carried out by the PSOs of Palermo

How important are the following organizational activities? (apart from the specific activity of the PSO)	1st 'very important'	4th or 'Not at all important'	No activity at all
Internal organization	30.0%	38.0%	0.7
Formation and refresher courses	29.1%	32.5%	5.3%
Planning and development of new services	18.5%	34.4%	11.3%
Communication	8.1%	60.4%	5.4%
Fund-raising	4.7%	47.3%	34.5%
Relationships with public administrations	4.7%	58.1%	6.1%
Recruitment of new members	4.7%	66.9%	12.8%
Relationships with other PSOs or third sector organizations	2.0%	67.3%	15.6%
Relationships with private companies	0.7%	54.1%	43.9%

Respondents: from 147 to 151 PSOs. Non respondents: from 4 to 8 PSOs, depending on the items. The total is less than 100, because the 2nd and the 3rd choice do not appear in this table.

The Problems of PSOs in Carrying Out their Communication Activities

Many PSOs in Palermo face a lot of problems in carrying out their communication activities. As you can see from table 2, most of the difficulties come from the mass media indifference, both at the national level (70%) and at the local one (62.5%): almost two out of three PSOs report that mass media indifference to their efforts is the first problem. The scarcity of resources, especially *tools* (67.1%), is considered another great problem.

Table 2: The problems of PSOs in Palermo in their communication activities

Which of the following problems is more frequently faced by a PSO in its communication activities?	1st 'Much '+ 'Enough'	4th 'Less' + 'None'
Indifference of the national mass media	70.0%	30.0%
Scarcity of resources: tools	67.1%	32.8%
Indifference of the local mass media	62.5%	37.5%
Scarcity of resources: staff	37.0%	63.0%
Difficulties in the organization	23.4%	66.6%
Low importance of social communication	14.4%	85.6%
Other problems (only 2 respondents)	40.0%	60.0%

Respondents: from 154 to 155 PSOs. Non-respondents: 1 or 2 PSOs for the first six items, 153 for the last one (open question). The total is less than 100, because the 2nd and the 3rd choice do not appear in this table.

The Communication Office
Only one out of two PSOs has a communication office (48.7%). A third of them do not have an office of their own, but rather use the communication office of the central organization they belong to (16.1%). A fifth of them use both the central office and its own (9.7%). Only one out of five PSOs carries out the communication activities through its own office (20.6%).

The Communication Activities
The main communication activities developed by this minority of PSOs through their own offices are threefold:

- The spokesperson of the PSO manager (carried out by two out of three PSOs – 66.7%);

- Information to the mass media (carried out by seven out of ten PSOs –70.2%);

- Communication to users (carried out by four out of five PSOs – 82.5%).

The PSOs and the Use of the New Media
A large part of the PSOs in Palermo owns both old and new media technologies. The questionnaire contained a lot of questions about them: i.e. the frequency of their use, who were the persons that used them more frequently, and so on. As you can see in table 3, the more diffused new media technologies in PSOs' offices are the *computer* (*online pc:* 79.9%; *offline pc:* 75.3%), and the *video cassette recorder* (60.4%). The other listed new media technologies are present in less than a half of the PSOs offices: *digital video camera* (42.9%), *video projector* (35.7%), *digital photo camera* (39.6%), *DVD player* (29.2%), *digital TV* (*broadcasting TV:* 11.7%, *satellite TV:* 11%). More than seven PSOs out of ten use the *online pc* at least once a week (71.9%), and about 60% use the *offline pc* (62.4%) with a similar frequency; but the *video cassette recorder* is used weekly only by a quarter of PSOs (26.7%). Much fewer in numbers are the PSOs that use all the other new media at least weekly.

Table 3: The use of the new media technologies in the PSOs in Palermo

Which of the following new media is present in the PSO office? And how frequently is it used for the PSO activities?	This medium is present in…	This medium is used at least weekly by…
Computer offline	75.3%	62.4%
Computer online	79.9%	71.9%
Digital photo camera	39.6%	7.8%
Digital video camera	42.9%	6.5%
DVD player	29.2%	7.1%
Broadcasting digital television	11.7%	2.6%
Satellite digital television	11.0%	1.9%
Video projector	35.7%	5.8%
Video cassette recorder	60.4%	26.7%

Respondents: 154 PSOs. Non-respondents: 1.

Do the PSOS have some kind of communicative competence on online communication? The following data, collected in Palermo, seem to authorize mild optimism. Almost all the PSOs in Palermo have members who know how to use the new media technologies. Basically all PSOs have members who can use a *mobile phone* and an *online computer* (99.4%), an *offline computer* (94.2%) and a *video cassette recorder* (93.5%). The PSOs having members who can use a *video camera* and a *video projector* are respectively 85.2% and 76.7%. Only a 51.5% of them can use a *video mobile phone*. On the basis of these findings, I do hope that the *Portal* can be a welcomed initiative: a lot of PSOs have members with a basic communicative competence on the new technologies.

Some Remarks on these First Findings

During this research action the team and I have been collecting much other data: about, for instance, the internal communication of PSOs, or the accountability of their financial spending for social activities, and so on. All these first findings indicate that the PSOs operating in Palermo need to improve their communication and find new channels. The new technologies may help them for the following reasons:

a) ICTs require low costs of access and management;

b) ICTs have a structural homology with social networks, hence with the structure itself of the PSOs.

Therefore, in comparison with the mass media, the ICTs offer to PSOs a greater chance to communicate.

A Research Action to Provide ICT-Support for the Communication of the PSOs in Palermo

A Practical Answer to the Need to Communicate

As mentioned before, the 'third sector' is an important part of civil society; it is indeed the most organized part of it. The PSOs produce relational goods and increase the social capital, but they have little or no visibility in the public sphere. Therefore, the question is: What can be done to promote and support the PSOs in their efforts to communicate?

A practical answer comes from Palermo. On November 2003, four institutions of the civil society founded the *Telematic Portal for the 'third sector' communication in Palermo* in order to help the local PSOs to communicate better through the ICTs; they are:

- The Department of Social Sciences of the University of Palermo;

- The Com.Pu.Lab. – the Public Communication Laboratory of the University of Palermo;

- The Office for Social Communications of the Archdioceses of Palermo;

- The *Caritas* Office of the Archdioceses of Palermo.

The *Portal* is the most visible outcome of this assistance, carried out through a research action, i.e. a sociological survey, which transforms the phenomenon it observes.

The Telematic Portal for the Communication of the 'Third Sector' in Palermo[3]

The main features of this *Portal* are

a) up to November 2004, i.e. after almost one year from the beginning of this research action in Palermo, more than 130 PSOs have asked and obtained to be included in the *Portal*. Nearly every single day other PSOs are identified in the metropolitan area and invited to join the *Portal*;

b) one webmaster and 12 'accounts' (see 3.3) are working together with the managers of the PSOs in order to provide each organization with its own web page on the *Portal*. Therefore, the *Portal* is currently hosting in the section 'Organizations' more than 130 web pages, each one of them presenting the following five communication elements:

the PSOs' logo;
the PSOs' location on the digital city map;
some digital photos introducing the PSOs' activities;
a selection of self-produced papers and documents;
the PSOs' communication plan.

In addition to the section 'Organizations', the *Portal* consists out of nine other sections, the most important being

- 'Third Sector', which gathers papers and documents produced by academic researchers all over Europe;

- 'Forum', a virtual exchange area where every PSO operating in Palermo can share ideas and suggestions about the topics proposed, such as, for example, 'The relationships between the 'third sector' and the City of Palermo';

- 'Chat line', a virtual room in which the volunteers can talk about their experiences;

- 'Informations', where each PSO can find news and legal advice about the proclamations issued by the Mayor of Palermo for provisions on social activities;

- 'Links' gives an easy access to the national centres and institutions operating in the 'third sector';

- 'Searching/Offering pro-social work' is a space where the offer and the search for pro-social work can meet.

A New Type of Voluntary Action: The Account of the 'Third Sector'
As advertisers know, the account is a well-known role in the organization of an agency: he or she is a person who takes care of the client, who helps him/her to define aims and targets, and who ultimately realizes the advertising campaign.

The *Portal* adopted this role from the for-profit sector and played it into the no-profit field. Therefore, the *account of communication* is a voluntary agent who helps the members of the PSOs in Palermo to communicate better both with other PSOs and with the civil society environment, especially the local authorities.

Therefore, this new interface of communication (the *Portal*) is not simply a new technological channel, but it is also a way to spread the culture of communication and reinforce the social ties both within the PSOs, and between them and the metropolitan area of Palermo. Moreover, the *Portal* gives visibility to the 'third sector' in the public sphere offering it new types of virtual presence and public debate.

Further Developments through Three Socio-Communicative Tools
As I mentioned above, the data about each PSO operating in Palermo were collected through three sociological questionnaires: the first two, exploring the culture of the 'third sector' and the organizational aspects of PSOs' activities, had already been used during the national survey of 2001–2002.

In this chapter, I have described some of the first data collected through a third questionnaire – a new one, tested by my research team at the University of Palermo – exploring the communication activities of the PSOs in Palermo.

In the near future it is my intention to extract new knowledge from this data; for instance, by comparing the national findings with the local ones, I hope to generate the profile of both the volunteers and the PSOs in Palermo.

Moreover, new qualitative data are going to be collected through the study of the activities connected to the 'Forum' or to other virtual presences in the *Portal*. Briefly, this interface of communication is a yard of experiments, a continuous challenge for the sociological imagination ...

An Interim Conclusion

In this chapter, I have presented the first findings obtained by a research action promoted by the University of Palermo within a larger project, which includes seven Italian universities working on the 'third sector' and the social capital in Italy. The research action promoted by the University of Palermo consists of a survey on all the PSOs operating in the capital of Sicily, and is aimed at helping the PSOs in their efforts to communicate better. For this purpose, a web interface has been implemented: the *Telematic portal for the communication of the 'third sector' in Palermo*. The research action is still going on, and other findings will arrive when a second survey is completed in spring 2005. On that occasion, the most important question will be answered, that is whether this research action will have succeeded in its main goal: the improvement of the PSOs' communication through ICTs, on the one hand, and the empowerment of local civil society, on the other.

Notes

1   Palermo is the capital of Sicily, the largest island in the Mediterranean Sea. With about 700,000 inhabitants, Palermo is the fifth city in Italy, after Milano, Roma, Torino and Napoli. Sicily has about 5 million inhabitants, and it is the third Italian region for population.

2   The University of Bologna (chair – North Italy); the Catholic University of Milano (North Italy); the University of Verona (North Italy); the University of Trento (North Italy); the University of Molise (Middle Italy); the University of Palermo (South Italy).

3   You can visit the *Portal* by clicking on the following url: http://www.terzosettorepalermo.it. The *Portal* has been officially presented to the civil society of Palermo on October 21, 2004, during an academic symposium on the theme: *The communication of the 'third sector' in Palermo and in the emerging net society*.

References

Bauman, Z. 2000. *Liquid Modernity*. London: Polity Press.

Beck, U., Giddens, A., Lasch, S. 1994. *Reflexive Modernization*. London: Polity Press.

Colozzi, I. 1997. 'Società civile e terzo settore', pp. 123–158 in Donati P. (ed.) *La società civile in Italia*. Milano: Franco Angeli.

Colozzi, I., Bassi, A. 1995. *Una solidarietà efficiente. Il terzo settore e le organizzazioni di volontariato*. Firenze: La Nuova Italia Scientifica.

Corbetta, P. 1999. *Metodologia e tecniche della ricerca sociale*. Bologna: Il Mulino; re-published in 2003 in 4 vols.: La ricerca sociale: metodologia e tecniche. Bologna: Il Mulino.

Donati, P. 1993b. *Teoria relazionale della società*. Milano: Franco Angeli.

Donati, P. 1997. *Pensiero sociale cristiano e società post-moderna*. Roma: Ave.

Donati, P., Colozzi, I. (eds.)1997. *Giovani e generazioni. Quando si cresce in una società eticamente neutra*. Bologna: Il Mulino.

Donati, P., Colozzi, I. (eds.) 2001. *Generare 'il civile': nuove esperienze nella società italiana*. Bologna: Il Mulino.

Donati, P., Colozzi, I. (eds.) 2002. *La cultura del civile in Italia: fra stato, mercato e privato sociale*. Bologna: Il Mulino.

Donati, P., Colozzi, I. (eds.) 2004. *Il Terzo Settore in Italia. Culture e pratiche*. Franco Angeli: Milano.

Frisanco, R., Trasatti, S., Volterrani, A. (eds.) 2000. *La voce del volontariato. Indagine nazionale su organizzazioni di volontariato e comunicazione*. Roma: Fivol.

Giddens, A. 1990. *The Consequences of Modernity*. Cambridge: Polity Press.

Glock, C. Y., Stark, R. 1966. *Religion and Society in Tension*. Chicago: McNally.

Jameson, F. 1984. *Postmodernism, or, the Cultural Logic of the Late Capitalism*. Durham (North Carolina, USA): Duke University Press.

Lyotard, J.-F. 1979. *La condition postmoderne*. Paris: Les Éditions de Minuit.

Martelli, S. 1999. *Sociologia dei processi culturali. Lineamenti e tendenze*. Brescia: La Scuola.

Martelli, S. 2004. 'Religione e comunicazione. Orientamenti di valore e uso dei media nel Terzo Settore italiano', pp. 227–256 in Donati P., Colozzi I. (eds.) *Il Terzo Settore in Italia. Culture e pratiche*. Milano: Franco Angeli.

6: What Price the Information Society? A Candidate Country Perspective within the Context of the EU's Information Society Policies

MIYASE CHRISTENSEN

Introduction

The 1990s, the years preceding WSIS, were marked by a number of radical initiatives implemented in order to bring ICT regimes increasingly outside of the national domain. The 1990s also witnessed vigorous European Union and United States telecommunications and IS policies.

This chapter, which is built on my recent case study of the EU's IST (Information Society Technologies) policies vis-à-vis candidate countries (Goktepeli, 2003), is an attempt to provide an insight into the current efforts to internationalize questions such as digital divide and communication rights from a country-specific perspective. In this regard, the focus remains on the discussion of the Turkish experience with regional IST policies, while I link my findings and arguments with the vision put forth through WSIS. The point of departure for this case study of Turkey is the contention that telecom infrastructure and the social shaping of national policy constitute the building blocks for the emergence of an IS (Information Society).[1] In other words, not only does the telecommunications infrastructure constitute the material basis for an IS, but the nature of and stakes around the infrastructure in a given context ultimately determine the nature of the IS or, preferably, a *communication society*' (Ó Siochrú, 2004) to emerge in that context. As Murdock (2004: 22–23) emphasizes:

> *Media scholars have tended to ignore the analysis of networks. For most, telecommunications policy, a long-standing and extensive area of research and debate, has remained a far away enclave of which they know little and cared to know less. In a context where popular telecommunications traffic was monopolized by voice telephony from fixed point and access was underwritten by principles of universal service this did not matter much. But in a commercialized communications environment where telecommunications links carry the full range of expressive forms, from images to video and music, the political economy of connectivity is increasingly central to a full analysis of the social organization of access and use.*

It is with these issues in mind that issues of telecommunications infrastructure and information society are addressed in relation to each other in this chapter. At their current stage, telecom policy and IS regimes in Turkey (a candidate to the EU) are shaped first and foremost by the binding policies of the EU and Turkey's own national power geometry, while the impact of a newly thriving civil society upon Turkish policy-making remains minimal. The findings presented in this chapter – apart from the discussion of WSIS – are based on policy analysis and personal interviews with over 35 stakeholders from Turkey and the EU. The interviews reveal useful insights into the web of power relations and personal/institutional conflicts otherwise missing from traditional policy analysis.[2]

The European Way

The current IST landscape in the EU region needs to be understood within the framework of the global developments in telecommunications over the last two decades. As the information economy has been expanding globally since the early 1990s, it has also enforced a new kind of global structure within which telecom flows take place. As Wilson (1992: 355) points out:

> *The transformation of the telecommunications industry from a regulated natural monopoly, which met the demands of the great majority of users, to a more competitive industry structure entailed a passage from the familiar to the unknown.*

One of the most remarkable milestones of this global move toward the 'unknown' has been the *Uruguay Round Final Act* embodying the results of the General Agreement on Trade and Tariffs (GATT) Uruguay Round of Multilateral Trade Negotiations, signed in 1995. This agreement, establishing the World Trade Organization (WTO) and aiming at the liberalization of trade around the world, was also approved by the Great National Assembly of Turkey on 26 January 1996.[3]

Parallel to the market-oriented logic of the international communication regime, as embodied within agreements such as GATT and NAFTA, the (infamous) *Bangemann Report* (1994) argued that, should the EU wish to catch up with the U.S. and the Asia-Pacific region, the development of the information sector in the EU should be based on private sector funding and commercial activity. The launch of the *National Information Infrastructure* (NII) initiative in the U.S. added additional impetus to European efforts to sustain competitiveness in the area of telecoms and ICTs. After Bangemann, there was a need for more concrete programs for action. In response – and parallel to the radical restructuring of the telecommunication sector at a time when most incumbent carriers in the European region (except for the UK) were state-owned monopolies operating with high telephony prices – the EU launched in the latter half of the 1990s region-wide *Information Society Action Plans* (ISAPs). Despite difficulties arising from the complex decision-making system of the EU, European telecommunications liberalization could be considered successful – at least in terms of reaching the desired economic goals (save for the later 3G disaster).

Throughout the 1990s, the European Union worked toward establishing a common regulatory regime in the telecom sector, which was not an easy task, considering the variety of political and institutional traditions that abide in the Union. As pointed out by Romano Prodi, then President of the European Commission, at the European Council held in Lisbon in March 2000, *'telecommunications liberalization in Europe is a success story'* (Cave & Prosperitti, 2001: 40). Data presented by Prodi supported this conclusion: between 1998 (when the EU entered a full competition regime in telecoms) and 1999, international call prices fell by an average of 40 percent; long distance prices by 30 percent; and, regional prices by 13 percent. Between 1998 and 2000 the total telecom services market grew by an estimated 12.6 percent, to 161 billion euros (Cave & Prosperitti, 2001).

The end of the 1990s marked a turning point in the field of European IST policy-making. At the 2000 Lisbon summit, which centred around information society issues, the heads of state of the 15 EU member countries set a very ambitious goal for Europe for the next decade: to become *'the most competitive and dynamic knowledge-based economy in the world'* (EC, 2000a). The EU recognized the need for Europe to further exploit the opportunities offered by the information society, and the *E-Europe Plan of Action* was officially launched on 20 June 2000. Establishing the infrastructure to enable economic and social activity in the new economy, or e-economy, was set as the primary goal of the *E-Europe* plan. Although the requirements of the initiative only applied to the (then) 15 members, the EU's regional goal to become the most dynamic and competitive knowledge economy in the world by 2010 inevitably held consequences for candidate members.

At the European Information Community Ministerial Conference, held in Warsaw in May 2000, the Central and East European Countries Information Community Joint High Level Committee (EU-CEEC JHLC) decided to form an action plan similar to *E-Europe* for the CEEC countries (Personal interview, 2002). The action plan was initially named the *E-Europe-like Action Plan*, and later changed into *E-Europe+*. The EU High Commission for Central and Eastern European Countries (CEEC) started the *E-Europe+* initiative following *E-Europe*. At its fifth meeting in October 2000, the Commission realized there were three other eligible countries outside of the CEEC group, and invited Malta, Cyprus and Turkey to join. The sixth meeting, held in March 2001, was the first attended by Turkey. The reasoning behind *E-Europe+*, as stated by a Turkish public sector official, was as follows: *'As the candidate members, we are all going to join the EU. Hence, whatever the requirements for E-Europe are, we should meet them and use E-Europe as a guide'* (Personal interview, 2002).

Clearly, the EU attributes a central role to the information society within the enlargement process. As noted at the European Ministerial Conference in June 2002: *'At this crucial moment in Europe's political development, we underline the importance of the Information Society in increasing social and cultural cohesion and in strengthening economic integration'* (EC, 2001). Accordingly, the desired outcome of the *E-Europe* and *E-Europe+* action plans is to allow member and candidate member countries to co-operate and to exchange information and experience so as to help the integration of Europe, and to avoid the further growth of the digital divide within the EU. The standardization of telecom regulation and the adoption of unified information society policies throughout the member and candidate member states, therefore, carry very significant implications for the future of the EU and for the accession of new members.[4] As stated in the EP's *Report on the 2000 Regular Report from the Commission on Turkey's Progress towards Accession*:

Telecommunications is a particularly important sector, since enlargement coincides with the advent of the Information Society [...] Almost all applicant countries are on course for full liberalization. It often happens that a regulatory authority has already been set up or decided on in principle, but its independence has yet to be secured. Substantial amounts of work still need to be done on infrastructure, services and adjustments to European standards. (EP, 2001: 42–43).

When one takes the EU's IST policy discourse at face value, regional information society policies seem to aim at the inclusion of large segments of the European population, and for good reason. But a lot of emphasis is placed on the problem of the digital divide, particularly in new member and candidate countries, for which the same medicine – liberalization of the telecommunications sector – is deemed a sufficient remedy.

As is evident from *E-Europe* (EC, 2000a; 2002c), *E-Europe+* (EC, 2000b) and other ISAP documents, the EU's IST policy rhetoric emphasizes, within a neoliberal framework, 'technological change' and 'economic imperatives' (i.e. the advantage of competition in terms of lowering prices in the public interest) as the reasons why increased *'marketization'* (Murdock, 2004) is needed. As discussed, a major driving force behind the EU Commission's decision to liberalize the sector has been the international trends towards expanding the scope of free trade in the area of telecommunications and, hence, the market imperative toward increased competition by penetrating into national markets. In line with this logic, and according to *E-Europe+*, the candidate countries had to take government – level action to meet the 14 targets by the end of 2003, targets which were gathered under four areas: (1) speeding up the formation of the information community basis; (2) cheaper, faster and more secure Internet; (3) investment in human resources; and (4), promoting Internet use.

In this regard, there is no basic difference between *E-Europe* and *E-Europe+*. The only difference is to be found in Article 0 of *E-Europe+*, which requires meeting the legislative criteria, *communitaire acquis*, of *E-Europe*. Harmonizing with the legislative framework of the EU is required in all other sectors including telecoms. *E-Europe+* also includes an article regarding the application of IS in environmental issues (Personal interview, 2002). Apart from these two basic differences, both *E-Europe* and *E-Europe+* rank-order the priorities in achieving IS as listed above: building infrastructure, training skilled people and carrying out implementation. What looks like a straightforward strategy, however, is quite problematic in the Turkish context. First, high levels of fixed telephony and online access is not a goal easily reachable within a liberalized telecom environment without a strong incumbent in monopoly position. Secondly, privatization of Turk Telekom (TT)[5] is on the Turkish government agenda but could have to happen in an already liberalized market, an anomaly. And third, the EU Commission wants universal service obligations enforced – which will scare off potential buyers of TT.

In short, there are significant gaps between old members, candidates and new members; and one size does not necessarily fit all. Until the middle of the 1990s, telecommunications services in Turkey were provided by the Post-Telegraph-Telephone (PTT)[6] administration, under the Ministry of Transport and Communications. A significant move towards liberalization and privatization of telecommunication services was made in 1994, when, in accordance with the new law, telecommunications services were separated, through corporatization, from the directorate general of PTT and transferred to Turk Telekom Co. Inc.[7] Since then, TT has operated as a public corporation but subject to public procurement law until recently – yet another anomaly which resulted in inefficiency. The gap between the member and the candidate members, in terms of the adoption of the *acquis*, is accounted for in *E-Europe+*, but with a rather simplistic approach:

> [...] E-Europe was launched at a time when the liberalization of the telecommunications sector was complete, the 1998 Telecoms acquis was adopted and implemented, and nearly all households had phone penetration in the EU. This is not the case in the candidate countries. Thus, a new objective 'acceleration of the work toward creating the fundamental building blocs of the information society' [Article 0] was added to E-Europe+, to address these three factors. (EC, 2000b: 18)

E-Europe+ affirms the determination of the candidate members to liberalize their telecoms markets: '*providing accessible communication services to all citizens is a fundamental necessity to prevent digital exclusion. Such services can be possible on the basis of a liberalized communication sector that operates within an efficient, competitive regulatory setting*' (EC, 2000b: 20). As articulated in policy discourse (and by Erkki Liikanen), the EU approach to ISTs is based on three pillars: 1) enhancing competition and investment in the market through a uniform legal and regulatory environment; 2) investing in R&D; and 3) promoting the use of the ISTs via *E-Europe*. Thus, liberalization of the telecoms market is seen as key to achieving the overall goals of these three pillars.

On this, one state sector informant from Turkey, who attended the Joint High Level Commission meetings, commented:

> At the E-Europe+ meetings, they wanted to put in a sentence like 'Internet prices will fall after privatization of telecoms in Turkey'. I objected to this, saying the Internet prices are already low in Turkey. Later, another country representative came up to me and said 'We privatized and it went up ... So, you are right.' ... This is one of the vague points in E-Europe+. It sets 'cheap Internet access for all' as a goal. But how is this going to happen? It suggests the strengthening of the Internet backbone as the solution. But Turkey doesn't have a problem like that. The backbone is fine. The backbone in Turkey is one of the best examples in Europe. But the user can't feel it because we are connected to a huge water pipe through a thin little hose. Or it suggests that all schools are connected to the Internet. But how? With which resources, with the goal of what? What kind of service should be offered? (Personal interview, 2002)

In regard to communication among candidates on the adoption of common goals, the same informant remarked:

> At the E-Europe+ meetings, you can easily voice your concerns or raise objections, since the other candidate countries are experiencing similar problems. Hungary and the Czech Republic are well ahead, but there are countries like Bulgaria and

Romania. And since the EU is sympathetic toward social concerns, they understand. But if you say you won't liberalize or privatize altogether, that will create problems, of course. (Personal interview, 2002)

Apart from the question of suitability of the scope of *E-Europe+*, one major problem is the cost of the work carried out within the *E-Europe+ Action Plan*, which comes predominantly from the national budget, private sector investments, related programs, and from the European Investment Bank, European Bank of Reconstruction and Development and the World Bank through programs founded by the EU. Since *E-Europe+* is an initiative that came from the candidate countries themselves, there are no direct EU funds available to candidate countries. For the EU members, however, there are special funds and subsidies available to allow countries such as Portugal to catch up. One possibility for the candidates is to use the available funds for certain projects based on *E-Europe+*, defining certain goals as *'priority areas'* (Personal interview, 2002). In Turkey, the action plan was managed and co-ordinated jointly by the Ministry of Foreign Affairs, the European Union Secretary General, and the secretariat of TUBITAK (Turkish Scientific and Technical Research Council of Turkey). According to the current time line, Turkey has to meet all the criteria by the end of 2006. While determining these dates, the expiration date of the MEDA (Euro-Mediterranean Partnership) fund in 2006 was taken as a basis in order to benefit from this fund, which the EU has reserved for such projects.

The Turkish Response: From *E-Europe+* to *E-Turkey* to *E-Transformation*

In September 2001, the Deputy Undersecretary of the Prime Ministry launched the *E-Turkey* action plan in order to reach the national goals as identified in *E-Europe+*. In March and May 2002 progress reports were published: the former of little substance, the latter relatively comprehensive. The 1st Progress Report on E-Europe+, Contribution from Turkey describes Turkey's joining *E-Europe+* as:

> *The information society policy studies, initiatives and projects in Turkey have gained a new impetus after launch of eEurope+ in June 2001. The existing efforts to transform the society into the harmonized combination of a knowledge-based economy and value adding citizens found a common appreciation at all levels of public, private, and non-governmental sector [...] The outcome of the close cooperation by all stakeholders was the eTurkey initiative with the international dimension realized by eEurope+.* (sic) (Office of the Prime Minister, 2002: 1)

In the same document, it was stipulated, on the progress made to date in *'providing affordable communication services for everyone [Article 0]'* in Turkey, that

> *The completion of the privatization process will prepare the basis for the liberalization of the sector. Considering that currently TT supplies infrastructure services to ISPs as a monopoly, Internet access will become cheaper and faster after the completion of the privatization process.* (Office of the Prime Minister, 2002: 3)

Within *E-Turkey*, 13 project groups were formed. A head institution was picked for each of these groups. The first report quoted above outlined the project areas and provided national demographic data as to the situation of IS in Turkey. The 13 work groups and respective coordinators as outlined in this report were: Education and Human Resources; Infrastructure; Legal infrastructure; Standards; Security; E-commerce; Investments and Planning; Archives and Digital Storage; International Monitoring and *E-Europe+*; Special Projects; Assessment of Current Situation; National Coordination and Monitoring; and, Environment and Health. These work groups, most of which were headed by public institutions, assessed the current situation, projects, initiatives, and policy studies in their respective fields. To facilitate comparative assessment and monitoring between the member and the candidate member countries, the candidates agreed on using the same indicators adopted by the 15 EU members for the purposes of benchmarking. Furthermore, in order to develop a common methodology and approach in collecting and presenting the related demographic data, it was agreed that the respective institutions of the candidates will work in close cooperation with their counterparts in the member countries. (Office of the Prime Minister, 2002)

After its launch, serious studies were undertaken in the scope of *E-Turkey*, although with little practical outcome. One informant from a civil society organization who heads one of the *E-Turkey* projects remarked:

> *When you look at these reports, you realize that we have moved only an inch. The second report is not even a status report. We should learn a lesson from this: unless we institutionalize these studies and unless we produce more professional work rather than amateur stuff, we can't get anywhere. This was realized, but a bit late. Finally, the government issued a memo and decided to bring together these project groups under a central office.* (Personal interview, 2002)

With the memo (Genelge, 2002), the then Deputy PM, Devlet Bahceli of the Nationalist Party, was assigned as the head of the action plan. The future of *E-Turkey* looked bright for a short while in spring 2002, before the coalition government fell apart due to the Prime Minister's fragile health. One informant from the public sector, who is also a member of the EU High Commission for CEEC, Joint High Level Commission, suggested there was no political consciousness as to IS prior to *E-Europe+*. '*In that regard,*' he further commented:

> *the only advantage of joining the E-Europe+ has been the launch of E-Turkey. But the prime minister announced this at a glamorous event, said a mouthful and nothing has been done since then [...] they understood this as 'something we need to do otherwise they won't accept us'. They joined not because they care about creating an IS in Turkey, but because joining the EU is a non-negotiable national goal. They formed some work groups and all that. Was it meaningful? Certainly not [...] E-Turkey doesn't really signify a national policy and approach to 'e'.* (Personal interview, 2002)

Another public sector informant who attended a number of *E-Turkey* work group meetings offered similar observations:

As far as I observed, academia and social scientists were completely left out of the work groups they formed. The groups were made up of computer technician kind of people, of informatics division heads and others appointed by them. So the method for pursuing E-Turkey was wrong. They should have started by launching an action plan [...] They formed some ad hoc work groups. Turkey wasted all progress made up to that point. I joined the meetings of some of these groups, and quit later. When you said universal service, most of these people understood to mean the Universal Declaration of Human Rights! Nothing could come out of those meetings. (Personal interview, 2002)

Another major development in the spring of 2002 was the Informatics Council held through cooperation between the under-secretariat of the Prime Ministry and four sectoral civil society organizations (CSO). The work toward organizing the event started in September 2001 upon an invitation letter by the then Prime Minister Ecevit. A report was issued following the Council and the participants presented a list of their expectations. Although the Council was a one-time event, the CSOs agreed on meeting every three months in order to discuss progress achieved. However, the participating CSOs never divided the ten or eleven key issues between themselves, and never agreed on carrying out specific projects (Personal interview, 2002). One CSO informant who participated in the Council commented:

The next meeting is on August 10[8]. I guess we will just get together and chat, since nothing has been done since May. This is common practice in our sector. We get together and chat on a lot of issues. We can't even satisfy ourselves at the end, let alone accomplishing something. (Personal interview, 2002)

Shortly after this, and upon the announcement of early national elections in July, all efforts in all areas were halted until November 2002. The Justice and Development Party (AKP), a centre-right party with former Islamist aspirations, won the office with 34% of the vote. Due to the popularity of the IS rhetoric in 2001–2, and thanks to *E-Turkey* and the Informatics Council, all political parties included 'e-transformation' (i.e. taking steps to make changes in certain aspects of society, economy and governance in order to catch up with the information age) in their election manifestos.

By the end of 2002, the EU was not happy with Turkey's homework. According to the EU's *2002 Progress Report on Turkey*, the country had *'made little progress since the last Regular Report.'* The report highlighted a number of troubling facts: that competition for fixed voice telephony would not be implemented before January 2004; penetration rates of not more than 28% in the fixed mobile network; Internet and household cable television penetration rates of only 4% and 5% respectively; and, the partial implementation of universal service. Urgent liberalization of fixed voice telephony by no later than 2004 and transposing the updated telecommunications *acquis* were prescribed along with other recommendations (EC, 2002a: 105).

From the start, the ruling party AKP pursued a pro-EU policy, and, hence, gave priority to privatization and liberalization of the state sector, including telecoms and informatics. Shortly after taking office, the government introduced an *Urgent*

Action Plan (sic) to address problem areas in IST policy implementation and to better coordinate relatively disorganized efforts. The first implementation period of the *Urgent Action Plan* was completed in December 2003. Within the Plan, as part of the Public Management Reform Section, *E-Transformation Turkey* was declared as a high-priority project and the responsible institution was identified as the State Planning Organization (SPO), which is directly affiliated with the Prime Ministry. A *Short Term Action Plan*, which covers 2003–2004, was put into force to implement specific tasks in eight areas: strategy, e-education and human resources, e-health, e-commerce, standardization, infrastructure and information security, legislation and e-government. To coordinate implementation, a new unit, Information Society Department, was also established within SPO.

In addition to the common goals adopted via *E-Europe+*, *E-Transformation* also identified a number of social objectives as priority areas, such as *'mechanisms that facilitate the participation of citizens to decision-making process in the public domain via the use of ICTs,'* *'enhancing transparency and accountability for public management'* and *'putting into place good governance principles through increased usage of ICTs'* (Genelge, 2003). To increase participation and success, an Advisory Board of 41 members from public institutions, NGOs and universities was established. In February 2003, with a PM's Circular, a new body, the Executive Board, was established. The Board included five members: the Minister of State and Deputy PM, the Minister of Transport, the Minister of Industry and Trade, the Undersecretary of SPO (State Planning Organization) and the Chief Counsellor to the PM. In addition, heads of eight other related organizations can participate in the board meetings. These are: the Heads of the E-Transformation Turkey Project Advisory Board; TUBITAK; Telecommunications Authority; the CEO of Turk Telekom; TOBB (The Union of Chambers of Commerce, Industry, Maritime Trade and Commodity Exchanges of Turkey); TBV (Turkish Informatics Foundation); TBD (Turkish Informatics Association) and TUBISAD (Informatics Industrialists' and Businessmen's Association of Turkey). The Board is in charge of running the *E-Transformation* project and steering actions in *Short Term Action Plan*.

In the telecom field, TT''s monopoly position ended on 1 January 2004, and the market opened up to other operators. A new privatization decree for TT was also issued in November 2003, in which it was stipulated that 51% of TT's shares be privatized through block sale, with the remaining shares put on public offer. As deemed necessary within the *Short Term Action Plan*, a new telecommunication law is also being prepared by the TA in order to amend the current laws in areas such as interconnection, licensing, universal service and numbering so that they fall into line with the EU *Acquis*. The TA is also undertaking the completion of necessary legislation to cultivate competition in the areas of VoIP licenses, long-distance telephony, cable services, network provision, rights of way, local loop unbundling, co-location and facilities sharing, numbering, personal data protection and consumer rights. Universal service is defined as *'minimum service'* according to Law No 4502, Article 1, and it includes *'public pay-phones, emergency telecommunication services and telephone directory services.'* The *Short Term Action Plan* requires that the preparation of a Directive for Universal Service by the

Ministry of Transport introduce incentives, financial grants and other necessary mechanisms with a legal basis.[9]

Currently, following liberalization of the sector in January 2004, the market environment remains turbulent. Long-distance telephony licenses and broadband provision is the subject of fierce clashes between TT, the TA and the private sector actors. TT is accused of breaching competition rules by misusing its dominant market position – for example, currently only TT is licensed to provide broadband services – and the TA for operating in the interest of TT. Considering that *de facto* liberalization of the mobile and ISP markets took place in the mid-1990s, the inception of a national regulatory authority in 2001 is a very late development. Moreover, the TA had to inherit a large number of staff from the Ministry of Transport and DG Wireless, it lacked regulation-making experience, and monitoring of the board was highly politicized.

At the end of the day, *E-Transformation Turkey* continues, and in a much more organized manner compared to the national projects preceding it. Nevertheless, at the execution level, the project seems to have been monopolized by industry groups such as TUBISAD and TBV, and there appears to be little understanding of the social issues at stake on the part of the media – other than a handful of columnists. While the *Action Plan* itself underlines the importance of transparency, the manner in which the project is carried out is far from meeting this principle. First of all, the fact that 'civil society' is represented only by a number of industry groups at the Execution Board meetings, with no participation from grass-roots organizations and community activist groups at any level, is an anomaly. Secondly, the meeting minutes are not made available for public access or to the media. The public is essentially in a no-win situation given the fact that only a fraction of Turkish society have access to online information; conventional media, for the most part, are turning a blind eye to the developments (despite the fact that the stakes are very high for industry groups and public sector, who want their share from the domestic IST pie); and, finally, the absence of any meaningful community-level activism. One of the two pillars missing in the Turkish equation, therefore, is the role of the press as auditor. The second is civil society organizations in the real sense of the term – not sectoral think tanks.

In addition, another adversity from an economic point of view appears to be lack of mutual trust between the public and private sectors. Currently, like the private sector, the state is in a process of restructuring in Turkey. However, both the private sector and sectoral CSOs tended to see the state as *'inefficient, visionless, bulky, clumsy, unjust and crooked.'*[10] The following comment by a private sector informant is emblematic of the sector's attitude toward the government. In response to my question regarding his view about E-Government applications in Turkey, he remarked:

> *Well, they need to start from a-government, the ABC of government, I mean, before e-government. The mentality of the state should change above all. Also, as long as there isn't enough number of users, what difference would it make even if they come up with the best e-government of the world?* (Personal interview, 2003)

A number of public and civil sector informants, on the other hand, commonly defined the private sector as *'greedy, lazy, simple-minded and shady.'* Thus, efficiency of the regulatory authority has an even greater importance in such an environment of mutual distrust, although particularly at the beginning the TA was far from providing satisfactory regulation and intervention. In the end, the TA was perceived as an extension of the state, and thus, it is seen as everything that is ascribed to the latter.

Where Does WSIS Leave Us? A Tale of Two Visions

During an interview with a private sector representative from Turkey on the suitability of the EU policy agenda, a revealing anecdote was offered which highlighted the potential chasm between the intent and actual application of policy:

> *ATM machines arrived quite early in Turkey. In fact, Turkey acts rapidly in receiving and using new technology. And it wasn't because people demanded it, but those who brought the technology realized that they would be of use in many ways and therefore they were needed. The first ones opened on Istiklal Street [a major street in Istanbul]. For the first 6 months, they were available only between work hours during the day. Only later they realized that this technology was meant to be used otherwise! Likewise, adopting the EU standards in a similar fashion is not going to solve anything.* (Personal interview, 2003)

Ultimately, the ways in which policies are implemented are determined by social, economic and cultural factors. In that regard, there are two realities involved in the current global, regional and national levels of IS-regimes: what is really needed and what economic actors desire. The official discourse of WSIS, like that of the EU Commission, was geared toward justifying that the latter is really what human society needs at this point in time.

Despite the problematic issues inherent in the WSIS initiative, the novelty and significance of the program stems from the fact that WSIS was the first international event bringing together multi-stakeholders-governments, civil society, private interest groups and bureaucrats – from all over the world to reflect on the future of IS from people-centered, communication/human rights perspectives: perspectives which are lacking in current national and supranational policies. So, is there hope for optimism? Yes, since the initiative, even if not in the form of any binding resolution for global, regional and national governing bodies, has opened up a discursive space for civil society organizations to vocalize their position on international IS governance and communication rights as major actors – through participation in the WSIS process itself and parallel alternative campaigns such as CRIS. These are spheres otherwise dominated by the techno-deterministic rhetoric of WTO, WB, and the regional/national power blocs such as the EU, U.S. and Japan.

If the aim of the summit was to create a 'common vision of IS' endorsed by the participating governments, then the event, from the point of view of the organizers', could be seen as a success. The *Declaration of Principles* (WSIS, 2003a) is all about common denominators such as equal opportunities in the digital age and cooperation among all stakeholders. Yet, because two potentially thorny issues – the fact that contributions into the Digital Solidarity Fund remain voluntary, and the question of Internet governance – were pushed forward to the second phase in Tunis in 2005, it is questionable to say that the summit bore fruit as it failed to solve problems identified within its own official agenda.

Ó Siochrú (2004: 203) suggests that there were two summits at WSIS: one of the Information Society ...

> ... the summit of information, telecommunication, the Internet, the 'digital' divide, and ultimately the neoliberal model of development, exposing its limits even as it strained to plead its relevance. The other was the summit on a knowledge and communication society, full of contradictions, ideas still in formation, but nevertheless beginning to perceive new potentials and possibilities. Each has its own distinct history. But only one has a future.

Ó Siochrú traces the lineage of 'information society' back to the 1970s, to the futurist visions put forth by Bell (1973) and Porat (1976) and sees the techno-deterministic approaches of the Bangemann Report (EC, 1994) and the later EU policy discourse as products of these visions. 'Communication society', he argues, is directly linked with the New World Information and Communication Order (NWICO), spearheaded by the Non-Aligned Movement (NAM) of UN countries during the second half of the 1970s and the first half of the 1980s (Ó Siochrú, 2004: 204–210). At the summit, these two visions materialized into two distinct documents: *Declarations of Principles* (WSIS, 2003a) (and the accompanying *Plan of Action* (WSIS, 2003b)), and the *Civil Society Declaration* (Civil Society Plenary, 2003).

Ó Siochrú's conceptual delineation of the two ideological realms and different stakes (global economic goals vs. human needs), which exist in stark contrast to each other (not only in terms of their genealogical trajectories but also in terms of the kind of future they envisage), is useful. It helps to sift through the pros and cons of 'information age' obscured by the grandiloquent policy discourse of regional and national governing bodies, and it also subverts the hollow significance attested to it within the neoliberal agenda. The creation and promotion of the 'information society' in Europe as elsewhere has been more an economic imposition, a forced effort, than a genuine development. As Calebrese and Burgelman (1999: 5) observed a few years back:

> For the past several years, a small industry dedicated to futuristic speculation and argument about the idea of the information society has existed, as is perfectly illustrated by the European Union's information society policies (and, similarly, in the United States and elsewhere). We are told that the evolution toward an information society is absolutely essential to improve the way things are and to allow us to be better citizens. The end of the cold war can also explain the success of this new discourse, when a particular clash of ideas ceased and a new mythology became necessary to mobilize society around the aims of capitalism.

More recently, Preston (2003: 51) notes that:

> we may note certain semantic shifts and genuflections towards a 'social Europe' agenda within recent spate of eEurope policy reports. But these seem little more than occasional rhetorical gestures in the midst of policy concepts and practices that are fundamentally embedded in the neoliberal ideology which celebrates a 'market-driven' information society and which privileges consumer identities and roles over those of citizenship.

Yet, while 'communication society' offers a much broader scope to change things in the right direction, as well as a lingual convenience for identifying ideological differences, such theoretical delineations do not translate into any difference in the current policy practices, particularly in countries bound by the EU policy agenda such as Turkey. If the aim is to bring the information society realm into the domain of communication society – and in the *Civil Society Declaration* the term 'information and communication societies' is used consciously – aspects of infrastructure governance, currently at the discretion of neoliberal policy-making and implementation, should be given priority in Tunis in order to *'build information and communication societies that are people centred, inclusive and equitable,'* (Civil Society Plenary, 2003) as underlined in the *Civil Society Declaration*. In that regard, solutions *within* the IS rhetoric first is much crucial. But this is not to suggest that the declaration does not address the importance of infrastructure. On the contrary, it does mention, at various levels, the importance of civil society and end-user participation in shaping technologies, and calls for financial support for sustainable e-development. It also argues for reforming international arrangements to augment network interconnections, frequency allocations, to ensure free trade, open public domain, protection of human rights, consumer safety and personal privacy, social and cultural diversity, and the prevention of the concentration of market power in ICT and mass media industries, all of which ultimately have to do with infrastructure. The declaration's recommendation to initiate public interest-oriented monitoring of intergovernmental and self-governance bodies is a step in the right direction, and this agenda needs to be pursued more vigorously in Tunis. In addition, civil society lobbies should gain representation at international policy-making bodies, such as the EU Commission, to influence and monitor policy output and implementation – easier said than done.

The other vision, as embodied in the official document output of the summit, is much more user-friendly as it does not require any change in current policy-making, particularly in the developed West. Before the summit, Commissioner Liikanen affirmed that the WSIS would provide the EU with the opportunity to point out what it considers key drivers for the IS: constant interaction with policy-making, regulation and technological development (EC 2003). In the *July 14 2004 Commission Communication* issued by the EU Commission, Commissioner Rehn confirmed that, *'the EU Commission is committed to continuing the road-map set out by the WSIS last year. We need to focus on bridging the digital divide and work to ensure access to the* information society for all *so that we have concrete deliverables at the next Summit in Tunis next year,'* (EC 2004; emphasis added). Based on this standpoint, the Commission proposed to implement the *Action Plan* around three axes: 1) creating an enabling environment based on eStrategies at all levels and on pro-competitive legal and regulatory frameworks that encourage investment and innovation; 2) showing applications that work in the areas of eGovernment, eLearning, eHealth and eBusiness; 3) paying special attention to the research dimension of IS. In addition, need for action for least developed countries was also addressed in the same communication.

Similarly, Turkish participation at the summit did not go beyond the official national agenda pursued to date. Turkey was represented by a delegate headed by the Minister of Transport, comprising of individuals from SPO (State Planning Organization), TA (Telecommunications Authority), the Ministry of Foreign Affairs and the Turkish Embassy to the UN. In his statement at the summit, the Minister of Transport, Mr. Yildirim, underlined the importance Turkey gives to building a global information society and pointed to the problem of global economic disparity as well as the question of a digital divide (SPO, 2004b). He called for contributions from developed countries for social-development projects in developing countries and suggested that intellectual property and patent policies be revised for the benefit of developing countries. Finally, the minister drew attention to the need for *'public-private partnerships and cooperation among governments on the one hand and private sector and the civil society on the other'* to implement policies to bridge the digital divide. The Minister concluded his statement by noting that

> *Freedom as access to information and knowledge is the cornerstone in transforming the world into an Information Society. As a prerequisite for the democratic societies governed by the rule of law, the right to access to information and knowledge should be included among the fundamental rights and freedoms and be defined as such at constitutional level. In this context, we would like to stress that the sharing and dissemination of the global information and knowledge would also contribute to the development of the desired level of international solidarity and cooperation in combating the scourge of terrorism which has a global character, as we witnessed in the light of the recent wave of terrorist attacks in different parts of the world.* (SPO, 2004b: 2)

It is interesting that among all the benefits of right to access information and knowledge has to offer, the Minister chose to identify an immediate link between this and fighting terrorism through sharing information, which, more often than not, translates into a breach of privacy rights.

Conclusion

Only when it came to a point where it was necessary to maintain the macro-economic balance as part of the Maastricht Agreement did the EU countries pass painful market reforms, in areas such as telecoms and energy. The same thing is true for the Lisbon strategies, which initiated the *'most competitive knowledge-based market'* process. (And the 6th Framework Programme is a tool designed completely for the realization of the Lisbon strategy.) Despite the social goals later pursued within *E-Europe* and *E-Europe+*, a technology-centred political economy marks the communication field in the EU region, and the achievement of economic growth and prosperity in this milieu depends heavily on the success of national IS strategies to be pursued – or so we are made to believe.[11] The fact that countries such as Turkey, with relatively low penetration rates, are expected to catch up with an 'information revolution' that took decades in the West in a matter of a few years – and in liberalized environments without the economies of scale of telecom monopolies – is absurd, but a fact nonetheless. As one informant, an EU representative of a major trade union, remarked:

> *In the EU, what they try to accomplish with ISTs is to catch up with the U.S. and to turn IS into something that directly lowers the costs and budget deficits for the government, that makes the expenditure more efficient and that turns the balance sheets of businesses in the positive direction, and something that materially produces surplus value, something that contributes into the economy. In other words, the purpose is not to win the Nobel Prize. It is to make money.* (Personal interview, 2002)

This pattern of policy-making in the EU raises many questions concerning governance in the EU region. First of all, the Commission is not an elected body. Although it draws its authority and legitimacy from national governments who concede to supranational governance, based on the fact that EU IST policies have been pursued, from the start, with an industry-pushed techno-deterministic rationalism, it is fair to suggest that the 'social' comes second to the 'market'. Significant reference is made to the participation of 'citizens' and the strengthening of 'democracy' in recent policy documents and ISAPs. Yet, as pointed out in various studies on EU communication policy, while the *E-Europe* initiative put a more 'human-centred', 'culturally and socially sensitive' face on EU activities in this area – particularly in comparison to the rigid neoliberalism of the telecom policies – information society policies of the EU, as Preston (2003: 49) puts it, are *'fundamentally framed, imagined and measured in terms of the maximum production and use of new ICTs'*. Parallel to that, the building blocks of democratic governance such as 'democratic participation', 'dialogue' and 'transparency' are commodified to increase demand and legitimize the market-oriented reforms. The treatment of ISTs as neutral, as is the case with the WSIS *Declaration of Principles,* and the lack of consideration of societal and cultural factors in the policy discourse contradict the EU's self-attested commitment to pluralism and diversity.[12] The massive amount of bureaucracy, which characterizes EU governance, also takes away from the transparency made possible by online access.

On the national level, while the global and regional context (in the form of binding agreements) provide the backdrop against which policy issues are approached in the EU region, Turkish policy-making follows a country-specific track: personal and institutional relations (conflicts, relationships of interest and rivalries) play a key role in shaping policy and regulatory output. As Williams and Edge (1996) suggest, technological change is patterned by the conditions of its creation. Policy-making is a key factor in shaping IST diffusion and use in the European context, and the way it is approached and implemented in Turkey might yield different penetration and use patterns than in other EU countries. The ways in which domestically driven forms of commodification respond to the regionally and globally driven forms of commodification is an important aspect of the political economy of communications today. In Turkey, domestic commodification took place within a statist environment and against the backdrop of a symbiotic power relationship between the state, military and the power elite. While the military has lost its prominence, for the most part, in influencing policy and legal decisions – thanks also to the reform packages Turkey passed over the few years to meet the EU criteria – nepotism both in public-private sector relations and within policy circles emerges as a major barrier to creating a market environment conducive to efficient competition. In relation to this, the concepts of 'public' and 'state sector' need to be redefined within their respective national contexts. The question of whether privatization and further marketization are desired is both an ideological and practical issue. However, due to a high level of corruption and inefficiency within the state sector, and to the adversities created by rent-seeking within and around TT, at this point a private monopoly appears to be a better, but an unlikely, alternative to the current structure in Turkey – given that sufficient institutional and operational transparency and efficient regulation are ensured.

Civil society also needs to be approached carefully and defined contextually. While it is referred to as a uniform social actor in policy, media and even in academic discourses, in Turkey, for example, it largely corresponds to sectoral non-governmental organizations and think tanks. This is not to suggest these groups do not count as civil society, but given that the term is laden with social roles and responsibilities ascribed to it (i.e. grass-roots movements and community activism) in popular and academic discourses alike, it is important to differentiate between sectoral lobbies and civil rights advocacy groups, which are much needed in transitional socio-economic realities of countries like Turkey. As a vital actor in democratic governance, the role of the CSOs in Turkey could be in terms of augmenting the heavy-handedness of state and private sector actors in policy-making.

In this chapter, I attempted to describe a national experience in the light of the existing forces of global/regional IS regimes and international aspirations expressed through the recent event of WSIS. In other words, I have tried to illustrate how, in Sassen's (1996) words, global processes materialize in national contexts. While a number of characteristics are distinctly national (i.e. the national history, certain ties between certain individuals and groups, etc.); many other aspects of the Turkish experience with telecoms restructuring (i.e. inexperience with independent regulation; lack of human and financial resources; institutional corruption; and the high amount of influence international organizations like the IMF and the WB exercise on domestic policies) are certainly regular fare in many other developing countries and new EU members. To benefit economically, Turkey needs to find her own means to support the national industry and to find areas where she can gain competitive advantage. One approach to the latter, as also suggested by some of the EU officials I interviewed, is that Turkey can be a good user: *'Why produce software or hardware or know-how when the others do it already?'* (Personal interview, 2002) This approach is prevalent in some circles in Turkey, too, and it carries important implications. Staying as a mere user-market increases the level of Turkey's dependency on European and global manufacturing, service and culture industries, and it also prevents the country from benefiting from the opportunities offered by the IST sector in the EU region. For the maximization of socio-economic and cultural benefits, the Turkish case points to the need for the Commission's consideration of transitional factors, and of a less techno-centric approach. Transferring electronic communication tools (e.g. e-government) alone does not mean anything, unless they are utilized in a meaningful way by large segments of the society.

Ultimately, given the influence of regional and global forces in policy-making, information society, an economic imperative, comes with a price tag. Whether it turns out to be worth the price, creates economic and social profits through right policies and use, or it turns out to be a waste, depends on the accuracy of the diagnosis and the effectiveness of the treatment. Increased convergence in the communication technologies and the audio-visual sectors poses even greater challenges for policy-makers, makes more ambiguous the relative roles of the social actors, and further complicates the process of governance. As one EU DG IS official put it, *'The development of technology is always, always faster, and this is why sometimes we would actually prefer to leave it to the market and to the industries'* (Personal interview, 2002). However, the dominance of market forces and the concentration of economic power in the spatial concentration of businesses can fragment the infrastructure *and* superstructure, which is counter-intuitive to the EU idea of further social unification. To reach the goal of a true 'communication society', policy-making that prioritizes social and cultural determinants – and research to identify these factors – as well as a socially adapted infrastructure, is a must, not a choice.

Notes

1   Here, I mean the taken-for-granted meaning of 'Information Society', a technocratic vision, as constructed within the international policy and media discourse, and do not suggest that it exists or is desirable as such. I comment on this point later in the chapter.

2   Interviews conducted between February 2002 - June 2003 in Turkey and Belgium.

3   Law: no 4067, The Ministry of Transport.

4   Turkey is adopting EU policies in accordance with the approach adopted in the following documents:

  • The Association Agreement between the European Community and Turkey (1963) and The 1970 additional protocol;

  • The Commission's communication on a European strategy for Turkey (4 March 1998);

  • The Commission's Regular Reports on Turkey's progress towards accession;

  • The Council Decision of 8 March 2001 on the principles, priorities, intermediate objectives and conditions contained in the Accession Partnership with Turkey (Turkish Ministry of Foreign Affairs).

5   TT is the incumbent operator in Turkey.

6   Founded in 1924.

7   Until 2001, some failed attempts took place toward privatizing Turk Telekom. Although the target date for finalizing the privatization of TT was the end of 2001, the tender did not go through due to lack of bidder interest and disagreements between the government and the military over the size of the stake at the time of the bidding. The recent depression in global stock markets, which hurt the telecom sector seriously, was among the factors that led to a lack of bidder interest in TT. Economic crisis and political instability at the time of the bidding made the telecom market in Turkey particularly risky for potential buyers and privatization was delayed until after the liberalization of the market in January 2004.

8   My interview with this informant took place on 23 July 2002.

9   Other major developments within the scope of these recent initiatives worth mentioning here briefly are: Electronic Signature Law 5070 issued on 23 January 2004 (to become effective on July 2004); Law Regarding Right of Information issues on 24 October 2003, to ensure transparency, openness and equality of public management; National Information Security Law and Personal Data Protection Law which were to be issued in 2004; and Secondary Legislation regarding Consumers' Protection Law to protect online consumers issued on 13 June 2004. Indicator data for the measurement of the success of implementation of the above goals is not available in Turkey most part, which remains a major problem. However, to cite some basic figures available in the recent *Progress Report: Contribution of Turkey to E-Europe+* (SPO, 2004a) dated January 2004: population: 71,251,000; PSTN penetration 26.3%; mobile phone penetration 39.3%; household income per month $610; Average cost of computer $600; percentage of people with PC 3.78%; percentage of people with Internet access NA; and Internet penetration 8.4%, all of which are well below EU averages.

10 All terms commonly used by the various stakeholders during the interviews.

11 The fact that the GSM operator formerly called ARIA, now Avea after merging with TT's own GSM operator Aycell, is the biggest foreign direct investment in Turkey through Telecom Italia is a serious indicator of the significance of telecom and IST sector in the general political economy of the country.

12 Not in the sense that the EU's IST policies directly discriminate against certain social and cultural groups, but in the sense that there are not multiple but one approach to the ISTs.

References

BBC News Online. 2003. 'Viewpoint: Summit will Create Common Vision'. Downloaded on 19 February 2004 from http://news.bbc.co.uk/1/hi/technology/3250862.stm.

Bell, D. 1973. *The Coming of the Post-Industrial Society: A Venture in Social Forecasting*. Harmondsworth: Penguin.

Calabrese, A., Burgelman, J. C. 1999. 'Introduction', pp. 1–13 in A. Calabrese and J. C. Burgelman (eds.) *Communication, Citizenship and Social Policy*. New York: Rowman & Littlefield.

Cave, M., Prosperitti, L. 2001. 'The Liberalization of European Telecommunications', pp. 39–77 in R. Crandall & M. Cave (eds.) *Telecommunications Liberalization on Two Sides of the Atlantic*. Washington, D.C.: AEI Brookings Joint Center for Regulatory Studies.

Civil Society Plenary. 2003. *Civil Society Declaration: Shaping Information Societies for Human Needs*, WSIS Civil Society Plenary, 8th of December. Downloaded on 21 January 2004 from http://wsis-online.net/smsi/file-storage/download/WSIS-CS-Decl-08Dec2003-eng1.htm?version_id=313554.

European Commission. 1994. *Europe and the Global Information Society: Recommendations to the European Council*, EU: Brussels (High-level group on the information society, aka Bangemann Report).

European Commission. 2000a. *E-Europe 2002*. Brussels: Information Society DG.

European Commission. 2000b. *E-Europe + A Cooperative Effort to Implement the Information Society in Europe*. Brussels.

European Commission. 2001. *European Ministerial Conference: Information Society and Connecting Europe*. Ministerial Conclusions. Ljubljana, 4 June 2002.

European Commission. 2002a. *Regular Report on Turkey's Progress toward Accession*, Brussels, 9th October 2002. Downloaded on 20 December 2002 from http://europa.eu.int/comm/enlargement/report2002/tu_en.pdf.

European Commission. 2002b. 'Mr Erkki Liikanen Member of the European Commission, responsible for Enterprise and the Information Society "Stimulating investment in European IT"', European Investment Forum Copenhagen, 5th November 2002. Downloaded on 20 December 2002 from http://europa.eu.int/rapid/pressReleasesAction.do?reference=SPEECH/02/541&format=HTML&aged=0&language=EN&guiLanguage=en.

European Commission. 2002c. *E-Europe 2005*. Brussels: Information Society DG.

European Commission. 2003. 'Mr Erkki Liikanen Member of the European Commission, responsible for Enterprise and the Information Society "Back on the Growth Path"', ITU TELECOM 2003 Conference Geneva, 12 October 2003. Downloaded on 21 January 2004 from http://europa.eu.int/rapid/pressReleasesAction.do?reference=SPEECH/03/460&format=HTML&aged=0&language=EN&guiLanguage=en.

European Commission. 2004. 'Towards a Global Partnership in the Information Society: Translating Principles into Actions', July 14 2004. Downloaded on 15 August 2004 from http://europa.eu.int/rapid/pressReleasesAction.do?reference=IP/04/898&format=HTML&aged=1&language=EN&guiLanguage=en.

European Parliament. 2001. *Report on the 2000 Regular Report from the Commission on Turkey's Progress towards Accession* (COM (2000) 713 _ C5 0613/2000 _ 2000/2014(COS)), 11 October. Downloaded from http://www.tbmm.gov.tr/ul_kom/kpk/Lamas.doc.

Genelge. 2002. E-Turkiye, B.02.0PPG.0.12-320-9259, 19 June 2002. Ankara: T.C Basbakanlik.

Genelge. 2003. *E-Dönüsüm Türkiye Projesi Kisa Dönem Eylem Plani*. Ankara: T.C Basbakanlik.

Goktepeli, M. 2003. *Telecommunications Policy and the Emerging Information Society in Turkey: An Analysis within the Context of the EU's Telecom and Information Society Policies*. Doctoral Dissertation, Austin: The University of Texas at Austin.

Murdock, G. 2004. 'Past the Posts: Rethinking Change, Retrieving Critique', *European Journal of Communication* 19(1): 19–38.

Ó Siochrú, S. 2004. 'Will the Real WSIS Please Stand up? The Historic Encounter of the "Information Society" and the "Communication Society"', *Gazette* 66(3–4): 203–224.

Office of the Prime Minister. 2002. *Information Society for All E-Turkey Initiative, I. Progress Report*, May 2002. Ankara, Turkey.

Porat, M. 1976. *The Information Economy*. Stanford, CA: Centre for Interdisciplinary Research.

Preston, P. 2003. 'European Union ICT Policies: Neglected Social and Cultural Dimensions', pp. 33–59 in J. Servaes (ed.) *The European Information Society: A Reality Check*. Bristol: Intellect Books.

Sassen, S. 1996. *Losing Control? Sovereignty in an Age of Globalization*. New York: Columbia University Press.

SPO-State Planning Organization IS Department. 2004a. *Contribution of Turkey to E-Europe + 2003 Progress Report*. Ankara, Turkey.

SPO-State Planning Organization IS Department. 2004b. 'Statement by His Excellency Binali Yildirim, Minister of Transport and Communications of the Republic of Turkey at the World Summit on the Information Society'. Downloaded on 7 April 2004 from http://www.bilgitoplumu.gov.tr.

Williams, R., Edge, D. 1996. 'The Social Shaping of Technology', pp. 284–299 in W. H. Dutton (ed.) *Information and Communication: Visions and Realities*. Oxford: Oxford University Press.

Wilson, G. W. 1992. 'Deregulating Telecommunications and the Problem of Natural Monopoly: A Critique of Economics in Telecommunications Policy', *Media, Culture & Society* 14: 343–368.

World Summit on the Information Society (WSIS). 2003a. *Declaration of Principles*, 03/Geneva/Doc/4-E, 10 December 2003. Downloaded on 20 December 2004 from http://www.mafhoum.com/press6/173T41.htm.

WSIS. 2003b. *Plan of Action*, 03/Geneva/Doc/5-E, 12 December 2003. Downloaded on 20 December 2004 from http://www.itu.int/dms_pub/itu-s/md/03/wsis/doc/S03-WSIS-DOC-0005!!MSW-E.doc.

7: Peer-to-Peer: From Technology to Politics

MICHEL BAUWENS

Introduction: Technology as both Embedding and Empowering Human Relationships

A New Template of Human Relationships?

This chapter is about 'a new template of human relationships'. First of all we should establish that such 'templates', general forms of human relationships, exist. For this we refer to the theory developed by Alan Page Fiske (1993), who argues that

> People use just four fundamental models for organizing most aspects of sociality most of the time in all cultures. These models are Communal Sharing, Authority Ranking, Equality Matching, and Market Pricing. Communal Sharing (CS) is a relationship in which people treat some dyad or group as equivalent and undifferentiated with respect to the social domain in question. Examples are people using a commons (CS with respect to utilization of the particular resource), people intensely in love (CS with respect to their social selves), people who 'ask not for whom the bell tolls, for it tolls for thee' (CS with respect to shared suffering and common well-being), or people who kill any member of an enemy group indiscriminately in retaliation for an attack (CS with respect to collective responsibility). In Authority Ranking (AR) people have asymmetric positions in a linear hierarchy in which subordinates defer, respect, and (perhaps) obey, while superiors take precedence and take pastoral responsibility for subordinates. Examples are military hierarchies (AR in decisions, control, and many other matters), ancestor worship (AR in offerings of filial piety and expectations of protection and enforcement of norms), monotheistic religious moralities (AR for the definition of right and wrong by commandments or will of God), social status systems such as class or ethnic rankings (AR with respect to social value of identities), and rankings such as sports team standings (AR with respect to prestige). AR relationships are based on perceptions of legitimate asymmetries, not coercive power; they are not inherently exploitative (although they may involve power or cause harm).
>
> In Equality Matching relationships people keep track of the balance or difference among participants and know what would be required to restore balance. Common manifestations are turn-taking, one-person one-vote elections, equal share distributions, and vengeance based on an-eye-for-an-eye, a-tooth-for-a-tooth. Examples include sports and games (EM with respect to the rules, procedures, equipment and terrain), baby-sitting coops (EM with respect to the exchange of child care), and restitution in-kind (EM with respect to righting a wrong). Market Pricing relationships are

oriented to socially meaningful ratios or rates such as prices, wages, interest, rents, tithes, or cost-benefit analyses. Money need not be the medium, and MP relationships need not be selfish, competitive, maximizing, or materialistic – any of the four models may exhibit any of these features. MP relationships are not necessarily individualistic; a family may be the CS or AR unit running a business that operates in an MP mode with respect to other enterprises. Examples are property that can be bought, sold, or treated as investment capital (land or objects as MP), marriages organized contractually or implicitly in terms of costs and benefits to the partners, prostitution (sex as MP), bureaucratic cost-effectiveness standards (resource allocation as MP), utilitarian judgments about the greatest good for the greatest number, or standards of equity in judging entitlements in proportion to contributions (two forms of morality as MP), considerations of 'spending time' efficiently, and estimates of expected kill ratios (aggression as MP). (Source: E-mail communication)

How is technology related to such types of sociality? We will argue in this chapter that technology both embeds social relationships and empowers them. We are neither defending a position of technological determinism, nor saying that technology simply reflects social or subjective structures, but that there are correlations and mutual influences. Our position is best reflected by those of philosopher of technology Andrew Feenberg, who argues that technology reflects in its very code, the contradictory social interests and world views. Technology is therefore a social construct reflecting deeply held epistemologies and ontologies.

A good example is the very structure of the Internet: originally commissioned by the military through their research programs (DARPA), it was designed as a decentralized network to survive nuclear wars, but it also went beyond that as is described by Janet Abate (1999). Reflecting the social values of the participating scientists at the end of the 1960s, it both reflected the political sensitivities of the era and the general values of peer-reviewed science based on open sharing of knowledge. Hence the network was designed to allow for a free flow of information and constant cooperation. At the same time, because of its very structure, and unlike previous forms of communication technology which were either one-to-one (the telephone) or one-to-many (print and mass media), it empowers many-to-many relationships and hence the autonomous networking of human groups. Significantly, e-mail was not planned by its conceptors but introduced by the early community of users.

The aim of this chapter is to describe this mutually influential relationship between the technological format, and the forms of human relationships that it reflects or empowers. To describe it, we will use the heuristic format described by Ken Wilber (2001) in his various books such as *A Theory of Everything*: indeed his four-quadrant descriptive scheme of the human lifeworld gives us a very useful descriptive tool. As a reminder, he says that every phenomenon has both an interior and exterior aspect (it has desires and motivations vs. it has/is a body in space), an individual and collective aspect (it has relative individual autonomy and agency, but, it is always already a part of a collective system). This gives us a quadrant system which distinguishes the field of the subject (the self, the 'I' perspective, the subjective), the field of the object (the body in space, the object, the 'it', the objective), the field of the intersubjective (the world view and immaterial aspects of systems and groups, the 'we' perspective), and finally the field of measurable systems, the interobjective (the 'its', political, economic, social, physical 'systems'). Note that since humans are characterized by the fact that they exteriorize the functions of the body and the brain in technological artefacts, that we will put technological developments in the quadrant of the 'object'. After undertaking our extensive survey of the emergence of P2P across these quadrants, we simplified Wilber's scheme even further and retained the following categories: 1) technology and the economy, 2) social organization and politics, 3) culture and spirituality.

As these respective fields have differentiated in modernity, and obtained a relative autonomy, we believe that if we can show that the proposed phenomena of peer-to-peer starts to appear consistently in the various fields, that we have a strong case that something is indeed brewing, and that it is indeed of a 'transformative' nature.

Definition and Scope of this Chapter

But what is peer-to-peer? Peer-to-peer is a specific form of a network, which lacks a centralized hierarchy, and in which the various nodes can take up any role depending on its capabilities and needs. Peer-to-peer is an 'egalitarian' network if you like, a form of 'distributive and cooperative intelligence'. Thus, intelligence can operate anywhere, and it lives and dies according to its capacities for cooperation and unified action. As we will see, it is related to Alan Page Fiske's typology in that it particularly 'reflects' and 'empowers' two particular forms of sociality: 'Equality Matching' and 'Communal Shareholding'.

Before reading the bulk of this chapter and its description of the emergence of peer-to-peer, it is important to know what I am saying, and more importantly, what I am not saying.

I am not arguing that technology in its P2P format inevitably creates a new type of society. Indeed I am fully aware that the current form of technology, despite its distributive and cooperative character, is embedded in an institutional framework which can make it function differently. The financial networks, which are globalized but nevertheless concentrated in key centers, is a good example. The use of Internet by Al Qaeda is another one. But, the seed of potentiality, which has already become in many respects an 'actuality', is there as well, and this is our focus. We believe that if a worldwide social movement would take up our concept, it would carry enormous power. Therefore, I am not saying that these developments will lead to political changes independent of human will and political and social struggle,

I am not painting a utopian future or saying P2P has only positive aspects. However, the pathologies and negative aspects of P2P are not within the scope of this particular chapter.

However, I am saying:

- Because of the social values that are embedded in the format, it enables and empowers particular social practices, such as 'Equality Matching', and 'Communal Shareholding', in particular.

- Because of such enablement, peer-to-peer can be a useful field of political promotion and struggle, especially for the social and political forces that favor such types of sociality.

- P2P can be a useful discourse, or language, that retranslates the emancipatory project in a way that is not only compatible with the new phase of cognitive capitalism, but also appeals to the new generation of youth, and additionally it can also find a linkage with the older forms of such socialites and the political and social movements and struggles that it produces.

- Because of the constraints of the space allocated to this article, I will restrain the scope of this article to the descriptive part.

Elsewhere, I have described the normative aspects of peer-to-peer, as well as its strategic aspects. The latter refers to the contradictory position of peer-to-peer as both the very infrastructure of 'cognitive capitalism', and as a practice that transcends and endangers its functioning. I have described three possible scenarios of 1) peaceful co-existence, 2) destruction of peer-to-peer in a context of information feudalism, 3) extension of the cooperative sphere until it becomes dominant. However, it is beyond the scope of this chapter.

The Emergence of P2P Across the Human Lifeworld: Technology and Economy

Peer-to-Peer as Technological Paradigm

Peer-to-peer is first of all a new technological paradigm for the organization of the information and communication infrastructure that is the very basis of our post-industrial economy. The Internet itself, as network of networks, is an expression of this paradigm. The early Internet was a pretty 'pure' peer-to-peer network, and it has now changed into being a network of unequal networks, many of them fully or partly walled, and with differential abilities. But nevertheless, it remains a network of networks, without centralization, and still functions as peer-to-peer, since no one is able to exclude participation.

Every node is capable of receiving and sending data. The peer-to-peer mode therefore makes eminent sense in terms of efficiency, as compared to the older models. It should be noted that, just as networks, peer-to-peer can come into many hybrid forms, in which various forms of hierarchy can still be embedded (as with the Internet, where all networks are not equal). If one surveys the technical literature, one realizes that there is no consistent definition of peer-to-peer, which is why we use a broader social definition. For example, the Web, though technically a client-server format, and though an unequal network with large and small publishers, socially still enables the free publication by any participant. Thus, according to the social definition, though imperfect, it is a peer-to-peer network.

As a technological format, peer-to-peer comes into two main forms. One is distributed computing, which takes advantage of the unused disk space and processing power at the edges of the Internet, i.e. all voluntary participating computers; and file-sharing, which distributes and places content, and sends the contact from computer to computer without having to pass to central servers.

Distributed computing is now considered to be the next step for the worldwide computing infrastructure, in the form of grid computing, which allows every computer to use its spare cycles to contribute to the functioning of the whole, thereby obviating the need for servers altogether. The telecommunication infrastructure itself is in the process of being converted to the Internet Protocol and the time is not all too far away where even voice will transit over such P2P networks. Last year, telecom experts have been able to read about developments such as Mesh Networks or Ad Hoc Networks, described in *The Economist*:

The mesh-networking approach, which is being pursued by several firms, does this in a particularly clever way. First, the neighborhood is 'seeded' by the installation of a 'neighborhood access point' (NAP) – a radio base – station connected to the Internet via a high-speed connection. Homes and offices within range of this NAP install antennas of their own, enabling them to access the Internet at high speed. Then comes the clever part. Each of those homes and offices can also act as a relay for other homes and offices beyond the range of the original NAP. As the mesh grows, each node communicates only with its neighbors, which pass Internet traffic back and forth from the NAP. It is thus possible to cover a large area quickly and cheaply. (The Economist, 2002)

Moreover, there is the worldwide development of Wireless LAN networks, by corporations on the one hand, but also by citizens installing such networks themselves, at very low cost.

In *Fortune* magazine, Stewart Alsop uncovered yet another aspect of the coming peer-to-peer age in technology, by pointing out that the current 'central server based' methods for interactive TV are woefully inadequate to match supply and demand:

Essentially, file-served television describes an Internet for video content. Anyone – from movie company to homeowner – could store video on his own hard disk and make it available for a price. Movie and television companies would have tons of hard disks with huge capacities, since they can afford to store everything they produce. Cable operators and satellite companies might have some hard disks to store the most popular content, since they can charge a premium for such stuff. And homeowners might have hard disks (possibly in the form of PVRs) that can be used as temporary storage for content that takes time to get or that they only want to rent – or permanent storage for what they've bought. (Alsop, 2002)

In general one could say that the main attractivity of peer-to-peer is that it will seamlessly marry the world of the Internet and the world of PCs. Originally, ordinary PC users who wanted to post content or services needed access to a server, which created inequality in access, but with true peer-to-peer file-sharing technologies, any PC user is enabled to do this.

P2P is superior because it places intelligence everywhere in the network; a total view of reality is no longer the privilege of the top of the hierarchy. Hence it enhances the collective intelligence of the entity adopting it, speeds up problem solving by mobilizing greater numbers, finding the answer faster by combining more perspectives and expertise. Almost in any technological endeavor, peer-to-peer is the solution to some kind of bottleneck created by the previous centralized form of organization.

Centralization is justified for two main reasons: 1) in a context of scarcity of intelligence, it makes sense to organize the flows; 2) it is a function of power and control. But in a context of the massive spread of computers, and of a mass intellectuality of an educated population, intelligence has become over-abundant, digital files can be reproduced at will at marginal cost, and such distribution often precludes the old styles of total control and organization. In such a context, centralization creates bottlenecks, and puts its users at a competitive disadvantage.

Peer-to-Peer as Distribution Mechanism

The last citation on the bottleneck concerning interactive TV points to yet another aspect of peer-to-peer: its incredible force as distribution mechanism. Indeed, the users of Personal Video Recorders such as TiVo are already using file-sharing methods that allow them to exchange programs via the Internet, and the model of TiVo is now emulated by almost all competitors and put as a standard feature of the new generation of cable modems. It is estimated that by 2004, half of American families will be equipped with it. But this is, of course, dwarfed by what is currently happening in the music world. Again the advantage here should be obvious, as in this mode of distribution, no centralizing force can play a role of command and control, and every node can have access to the totality of the distributed information.

The latest estimates say that:

> *Worldwide annual downloads, according to estimates from places like Webnoize, would indicate that the number of downloads – if you assume there are 10 songs on a CD – is something like five times the total number of CDs sold in the U.S. in a year, and one-and-a-half times the worldwide sales.* (Cave, 2002)

The original file-sharing systems, such as Napster, AudioGalaxy, and Kazaa, still used central servers or directories which could be tracked down and identified, and thus attacked in court, as indeed happened, thereby destroying these systems one by one. But today, the new wave of P2P systems avoid such central servers altogether. The most popular current system, an expression of the free software community, i.e. Gnutella, had over 10 million users in mid-2002, and as they are indeed distributed and untraceable, have been immune to legal challenge. Though the industry has used a variety of legal means to thwart the growth of file sharing, and even caused a dip in its uptake, as we write, usage is up again. Significantly, commercial forces, such as Apple iTunes/iPod, are adapting commercial versions (though with severe restrictions), and are in the process of convincing industry majors to adopt such a modified model.

But let us not forget that it will be very difficult to emulate the universal access, infinite flexibility in usage, and marginal distribution costs, of the existing file-sharing systems.

Peer-to-Peer as Production Method

P2P is not just the form of technology itself, but increasingly, it is a 'process of production', a way of organizing the way that immaterial products are produced (and distributed and 'consumed'). The first expression of this was the Free Software movement launched by Richard Stallman (2002). Expressed in the production of software such as GNU and its kernel Linux, tens of thousands of programmers are cooperatively producing the most valuable knowledge capital of the day, i.e. software. They are doing this in small groups that are seamlessly coordinated in the greater worldwide project, in true peer groups that have no traditional hierarchy. Eric Raymond's seminal essay/book *The Cathedral and The Bazaar* (2001) has explained in detail why such a mode of production is superior to its commercial variants.

Richard Stallman's Free Software movement is furthermore quite radical in its values and aims, and has developed legal devices such as Copyleft and the General Public License, which uses commercial law itself to prohibit any commercial and private usage of the software. Projects such as the Creative Commons initiated by Lawrence Lessig (2004), are extending the concept beyond software, to authorship in general.

Here is an explanation of the free software concept:

> *'Free software' is a matter of liberty, not price. To understand the concept, you should think of 'free' as in 'free speech,' not as in 'free beer.'*
> *Free software is a matter of the users' freedom to run, copy, distribute, study, change and improve the software. More precisely, it refers to four kinds of freedom, for the users of the software:*
> *The freedom to run the program, for any purpose (freedom 0).*
> *The freedom to study how the program works, and adapt it to your needs (freedom 1). Access to the source code is a precondition for this.*
> *The freedom to redistribute copies so you can help your neighbor (freedom 2).*
> *The freedom to improve the program, and release your improvements to the public, so that the whole community benefits. (freedom 3). Access to the source code is a precondition for this.* (Free Software Association, 2004)

Less radical, and perhaps more widespread because of this, is the Open Source movement launched by the above-mentioned Eric Raymond, which stipulates that the code has to be open for consultation and usage, but where there are restrictive rules and the property remains corporate. Together, even in a situation where the software world is dominated by the Microsoft monopoly, these two types of software have taken the world by storm. The dominant server of the Internet (Apache) is Open Source, but more and more governments and businesses are using it as well, including in mission-critical commercial applications. Many experts would agree that this software is more efficient than its commercial counterparts. What is lacking today is the spread of user-friendly interfaces, though the first open source interfaces are coming into existence and programs such as OpenOffice are beginning to be used.

Please also remember that peer-to-peer is in fact the extension of the methodology of the sciences, which have been based since 300 years on 'peer review'. Scientific progress is indeed beholden to the fact that scientists are accountable, in terms of the scientific validity of their work, to their peers, and not to their funders or bureaucratic managers. And the early founders of the Free Software movement were scientists from MIT, who exported their methodology from knowledge exchange to the production of software. In fact, MIT has published data showing that since a lot of research has been privatized in the U.S., the pace of innovation has in fact slowed down. Or simply compare the fact of how Netscape evolved when it was using Open Source methods and was supported by the whole Internet community, as compared to the almost static evolution of Internet Explorer, now that it is the property of Microsoft.

The methodologies initiated by the Free Software and Open Source movements are rapidly expanding into other fields; witness the movements such as the royalty-free music movement, the Open Hardware project (and the Simputer project in India), OpenTV and many much more of these type of cooperative initiatives.

I would like to offer an important historical analogy here. When the labor movement arose as an expression of the new industrial working class, it invented a series of new social practices, such as mutual aid societies, unions, and new ideologies. Today, when the class of knowledge workers is socially dominant in the West, is it a wonder that they also create new and innovative practices that exemplify their values of cooperative intellectual work?

And is it not particularly significant that the industry majors, who champion an economic system that claims to be the most efficient in terms of innovation, is putting all its energies in the stifling of technological innovation, much like the medieval guilds and nobility tried to stop the new practices of the early industrialists?

Peer-to-Peer in Manufacturing?

We would in fact like to go one step further and argue that peer-to-peer will probably become the dominant paradigm, not just in the production of immaterial goods such as software and music, but increasingly in the world of manufacturing as well. This has recently been argued by Steve Weber (2004), professor of political science at U.C. Berkeley, who maintains:

> that the open source community has built a mini-economy around the counterintuitive notion that the core property right in software code is the right to distribute, not to exclude. And it works! This is profound and has much broader implications for the property rights regimes that underpin other industries, from music and film to pharmaceuticals. Open source is transforming how we think about 'intellectual' products, creativity, cooperation, and ownership – issues that will, in turn, shape the kind of society, economy, and community we build in the digital era. (Publisher statement, e-mail communication)

Two recent examples should illustrate it. Lego Mindstorms is a new form of electronic Lego, which is not only produced by Lego, but where thousands of users are themselves creating new building blocks and software for it. The same happened with the Aibo, the artificial dog produced by Sony, which users started to hack, first opposed by Sony, but later with the agreement of the company. This makes a lot of sense, as indeed, it allows companies to externalize R&D costs and involve the community of consumers in the development of the product. This process is becoming generalized. Of course, work has always been cooperative (though also hierarchically organized), but in this case, what is remarkable is that the frontier between the inside and the outside is disappearing. This is in fact a general process of the Internet age, where the industry is moving away from mass production to one-to-one production or 'mass customization', but this is only possible when consumers become part and parcel of the real production process. If that is the case, then that of course gives rise to contradictions between the hierarchical control of the enterprise vs. the desires of the community of users-producers. It can also potentially give rise to new forms of social production, which bypass the corporate model altogether.

At a conference of Oekonux, the engineers of Volkswagen and Siemens who were present were adamant that the model of Open Sources was exportable to industry, and this is also the point of view of Steve Weber, in the above-mentioned book.

Some Preliminary Considerations

One has, of course, to ask oneself, why is this emergence happening, and I believe that the answer is clear. The complexity of the post-industrial age makes the command and control approaches, based on centralization, inoperable. Today, intelligence is indeed 'everywhere' and the organization of technology and work has to acknowledge that.

And more and more, we are indeed forced to conclude that peer-to-peer is indeed a more productive technology and way of organizing production than its hierarchical, commodity-based predecessors. This is of course most clear in the music industry, where the fluidity of music distribution via P2P is an order of magnitude greater, and at marginal cost, than the commodity-based physical distribution of CDs.

What is important is that peer-to-peer is a continuously offensive strategy, and implicitly creates a new public domain, and that industry is on the defensive.

Social Organisation and Politics

P2P is also emerging as the new way of organizing and conducting politics. The alter-globalization movement is emblematic for these developments:

- they are indeed organized as a network of networks,

- they intensively use the Internet for information and mobilization and mobile (including collective e-mail) for direction on the ground,

- their issues and concerns are global from the start,

- they purposely choose global venues and heavily mediated world events to publicize their opposition and proposals.

Here is a quote by Immanuel Wallerstein (2002, see also 2004), 'world system' theorist and historian, on the historic importance of Porto Alegre and its network approach to political struggle:

Sept. 11 seems to have slowed down the movement only momentarily. Secondly, the coalition has demonstrated that the new antisystemic strategy is feasible. What is this new strategy? To understand this clearly, one must remember what was the old strategy. The world's left in its multiple forms – Communist parties, social-democratic parties, national liberation movements – had argued for at least a hundred years (circa 1870–1970) that the only feasible strategy involved two key elements – creating a centralized organizational structure, and making the prime objective that of arriving at state power in one way or another. The movements promised that, once in state power, they could then change the world.

This strategy seemed to be very successful, in the sense that, by the 1960s, one or another of these three kinds of movements had managed to arrive at state power in most countries of the world. However, they manifestly had not been able to transform the world. This is what the world revolution of 1968 was about – the failure of the Old Left to transform the world. It led to 30 years of debate and experimentation about alternatives to the state-oriented strategy that seemed now to have been a failure. Porto Alegre is the enactment of the alternative. There is no centralized structure. Quite the contrary. Porto Alegre is a loose coalition of transnational, national, and local movements, with multiple priorities, who are united primarily in their opposition to the neoliberal world order. And these movements, for the most part, are not seeking state power, or if they are, they do not regard it as more than one tactic among others, and not the most important. (Wallerstein, 2002)

This analysis is confirmed by Michael Hardt, co-author of *Empire* (2001), the already classic analysis of globalization that is very influential in the more radical streams of the anti-globalization movement:

The traditional parties and centralized organizations have spokespeople who represent them and conduct their battles, but no one speaks for a network. How do you argue with a network? The movements organized within them do exert their power, but they do not proceed through oppositions. One of the basic characteristics of the network form is that no two nodes face each other in contradiction; rather, they are always triangulated by a third, and then a fourth, and then by an indefinite number of others in the web. This is one of the characteristics of the Seattle events that we have had the most trouble understanding: groups which we thought in objective contradiction to one another – environmentalists and trade unions, church groups and anarchists – were suddenly able to work together, in the context of the network of the multitude. The movements, to take a slightly different perspective, function something like a public sphere, in the sense that they can allow full expression of differences within the common context of open exchange. But that does not mean that networks are passive. They displace contradictions and operate instead a kind of alchemy, or rather a sea change, the flow of the movements transforming the traditional fixed positions; networks imposing their force through a kind of irresistible undertow. (Hardt, 2002)

Here is also a description by Miguel Benasayag (see Benasayag & Sztulwark, 2002) of the type of new organizational forms exemplified in Argentina:

> M.B. : *Les gens étaient dans la rue partout, mais il faut savoir quand même qu'il y a une spontanéité 'travaillée', pour dire ce concept là. Une spontanéité travaillée, cela ne veut pas dire qu'il y avait des groupes qui dirigeaient ou qui orchestraient ça, bien au contraire. Quand arrivaient des gens avec des bannières ou des drapeaux de groupes politiques, ils étaient très mal reçus à chaque coin de rue. Mais en revanche, une spontanéité 'travaillée' en ce sens que l'Argentine est 'lézardée' par des organisations de base, des organisations de quartier, de troc...*

> C.A. : *Lézardée, c'est un maillage?*

> M.B. : *Oui, c'est ça, il y a un maillage très serré des organisations qui ont créé beaucoup de lien social. Il y a des gens qui coupent les routes et qui font des assemblées permanentes pendant un mois, deux mois, des piqueteros. Il y a des gens qui occupent des terres ... Donc cette insurrection générale qui émerge en quelques minutes dans tout le pays, effectivement elle émerge et elle cristallise des trucs qui étaient déjà là. Donc c'est une spontanéité travaillée; c'est à dire que quand même il y a une conscience pratique, une conscience corporisée dans des organisations vraiment de base. C'est une rencontre du ras-le-bol, de l'indignation, de la colère populaire, une rencontre avec les organisations de base qui sont déjà sur le terrain. J'étais en Argentine quelques jours avant l'insurrection. Et il y avait partout partout des coupures de routes, des mini insurrections. Et ce qui s'est passé, c'est qu'il y a eu vraiment comme on dirait un saut qualitatif: les gens en quantité sortent dans la rue et y rencontrent les gens qui étaient déjà dans la rue depuis très longtemps en train de faire des choses. Et cela cristallise et permet de faire quelque chose d'irréversible.'* (Courant Alternatif, 2002)

What is significant is that the Argentinean demonstrators seemed to reject the whole political class, not just the established parties but also the left-wing radicals who wanted to speak for them and 'centralise their struggles', clearly opting for various forms of self-organization. So here, the often-decried anti-politics have a whole different context, not as a sign of apathy, but as a sign of rejection of hierarchical forms. Also related is the extraordinary rapid resurgence in Argentina of barter systems, based on the Local Exchange Trading Systems, which in a very short time succeeded in mobilizing hundreds of thousands of Argentineans. While the Argentine crisis is now less acute, and traditional politics is once again on the ascendant, many of the social practices described above are still being practiced.

A report from the Canadian Security Intelligence Service has paid particular attention to the innovative organizing methods of the alterglobalization protesters, and to their use of technology: Internet before and after the event and cell phones during the events. It concludes that with these innovations, established police powers have great difficulty to cope:

> *Cell phones constitute a basic means of communication and control, allowing protest organizers to employ the concepts of mobility and reserves and to move groups from*

place to place as needed. The mobility of demonstrators makes it difficult for law enforcement and security personnel to attempt to offset their opponents through the presence of overwhelming numbers. It is now necessary for security to be equally mobile, capable of readily deploying reserves, monitoring the communications of protesters, and, whenever possible, anticipating the intentions of the demonstrators. (E-mail communication)

Here's an example of P2P organizing at the extreme right, related to what is reportedly one the fastest growing radical religions today, the Odinists:

Today, the number of white racist activists, Aryan revolutionaries, is far greater than you would know by simply looking at traditional organizations. Revolutionaries today do not become members of an organization. They won't participate in a demonstration or a rally or give out their identity to a group that keeps their name on file, because they know that all these organizations are heavily monitored. Since the late 1990s, there has been a general shift away from these groups on the far right. This has also helped Odinism thrive. Odinists took the leaderless resistance concept *of [leading white supremacist ideologue] Louis Beam and worked on it, fleshed it out. They found a strategic position between the upper level of known leaders and propagandists,* and an underground of activists who do not affiliate as members, but engage instead in decentralized networking and small cells. *They do not shave their heads like traditional Skinheads or openly display swastikas.* (Southern Poverty Law Center, 2001 – my emphasis)

Culture & Spirituality

Peer-to-Peer in the Spiritual Field

Starting in the late 1980s arose a critical counter-movement against the feudal, authoritarian, patriarchal elements extant in the various world religions, but particularly as a reaction against the abusive practices generated by a number of 'spiritual masters' active in the West, but representing Eastern traditions. One such critique is expressed in *The Guru Papers* by the Kramers (1993), and in a critique of the hierarchical assumptions of Eastern spiritualities.

As a result, there has been the emergence of a great number of 'peer circles', which are based on peer-to-peer relationships, where a number of spiritual searchers, which consider themselves to be equals, collectively experiment and confront their experiences. This has been elaborated into a methodology by John Heron (1998) in his books on *Cooperative Inquiry* and *Sacred Science*, and also in the important new book by Jorge N. Ferrer (2001), *Revisioning Transpersonal Theory*:

Ferrer argues that spirituality must be emancipated from experientialism and perennialism. For Ferrer, the best way to do this is via his concept of a 'participatory turn'; *that is, to not limit spirituality as merely a personal, subjective experience, but to include interaction with others and the world at large. Finally, Ferrer posits that spirituality should not be universalized. That is, one should not strive to find the common*

thread that can link pluralism and universalism relationally. Instead, there should be
emphasis on plurality and a dialectic between universalism and pluralism. (Paulson,
2002 – my emphasis)

The above description is important because it also signals a shift to the use of peer-
to-peer, not just as a descriptive tool, but as a normative tool, reflecting a new set
of social demands, embedded in which is also a social critique of 'Authority
Matching' and 'Market Pricing' as it dominates fields of human endeavor.

A New Culture of Work and Being
Pekka Himanen (2002) has examined another cultural aspect of peer-to-peer,
based on his analysis of the work culture of the free software and hacker
communities, in his book about *The Hacker Ethic*. In this book, he compares the
Protestant work ethic defined by Max Weber (2001) is his classic *The Protestant*
Ethic and the Spirit of Capitalism, with the new mentality of hackers.
    A quote from the blurb:

> *Nearly a century ago, Max Weber articulated the animating spirit of the industrial*
> *age, the Protestant ethic. Now, Pekka Himanen – together with Linus Torvalds and*
> *Manuel Castells – articulates how hackers represent a new, opposing ethos for the*
> *information age. Underlying hackers' technical creations – such as the Internet and*
> *the personal computer, which have become symbols of our time – are the hacker val-*
> *ues that produced them and that challenge us all. These values promoted passionate*
> *and freely rhythmed work; the belief that individuals can create great things by join-*
> *ing forces in imaginative ways; and the need to maintain our existing ethical ideals,*
> *such as privacy and equality, in our new, increasingly technologized society.*
> (Himanen, 2002: cover)

This same aspect is discussed by Kris Roose on the website noosphere.cc, where
he distinguishes the 'secondary culture', described originally by Max Weber, where
one works, many times unpleasantly, to make a living and buy oneself pleasures,
and the tertiary culture, where the work itself becomes an expression of oneself
(the 'self-unfolding' process described by Stephan Merten of Oekonux.de, see
below) and a source of direct pleasure.
    In his book, Himanen first describes how that what the Calvinists and
Protestants actually did, was extending the work ethic of the Christian monasteries
to the whole of society, a process of 'Friday-ization'. In cognitive capitalism, this
process reaches its zenith, and he cogently argues how the popular Personal
Development ideologies promoted in the corporate world, are an extension and
extreme-ization of the Protestant work ethic, but adapted to the network world, and
made devoid of its ethics. This leads to the very unwelcome development of the
'Friday-ization of Sunday', so that the ethic of productivity and efficiency is
contaminating our personal and familial lives, which have become 'psychologically
unsustainable'. But he says, there is a counter-movement at work, a counter-ethic,
exemplified by the hackers (in the original meaning of the term, i.e. free software
programmers), where one finds the process of the 'Sunday-ization of Friday' taking
place. Indeed, work for them is a process of self-unfolding of creative interests, of

cooperative working and learning, of play, of intensive periods of 'flow', followed by extensive periods of rest and renewal. This culture, which is also in evidence in some creative industries, should be extended to the whole of industry, and this is in fact what is demanded by the new generations.

Richard Barbrook and other writers of *The Digital Artisans Manifesto* had already described some of the elements of this culture as well:

4. We will shape the new information technologies in our own interests. Although they were originally developed to reinforce hierarchical power, the full potential of the Net and computing can only be realized through our autonomous and creative labor. We will transform the machines of domination into the technologies of liberation.

9. For those of us who want to be truly creative in hypermedia and computing, the only practical solution is to become digital artisans. The rapid spread of personal computing and now the Net are the technological expressions of this desire for autonomous work. Escaping from the petty controls of the shop floor and the office, we can rediscover the individual independence enjoyed by craftspeople during proto-industrialism. We rejoice in the privilege of becoming digital artisans.

10. We create virtual artifacts for money and for fun. We work both in the money-commodity economy and in the gift economy of the Net. When we take a contract, we are happy to earn enough to pay for our necessities and luxuries through our labors as digital artisans. At the same time, we also enjoy exercising our abilities for our own amusement and for the wider community. Whether working for money or for fun, we always take pride in our craft skills. We take pleasure in pushing the cultural and technical limits as far forward as possible. We are the pioneers of the modern. (Barbrook & Schultz, 2002)

But hackers are not in fact the only ones exemplifying those values of working for passion, based on self-unfolding of one's creativity and desires, and in the context of peer-based relationships. A whole new generation of youngsters have shown to be ready for such social practices, as shown in the book by Andrew Ross (2001) *No-Collar*, where he coined the concept of the *'Industrialization of Bohemia'* and says these practices were exemplified for a short number of years in the dynamism of the Internet start-ups, before they were destroyed by the short-termism of their venture capital backers. We are in fact talking about new ways of feeling and being. We should note how the author also stresses the high human cost of such ways of working, when they clash with the contrary logic of for-profit management.

In our previous paragraph on peer-to-peer-based forms of political organizing, we quoted Miguel Benasayag, the philosopher who is going furthest in identifying new cultural substrata that makes P2P practices possible. (He has of course been influenced by the paradigmatic work of what we could call the 'founding P2P philosophers', Gilles Deleuze and Felix Guattari (1980), whose first chapter of their classic *Milles Plateaux* is dedicated to a description of the 'Rhizome', a complete peer-based network ...)

C'est pourquoi nous pensons que toute lutte contre le capitalisme qui se prétend globale et totalisante reste piégée dans la structure même du capitalisme qui est, justement, la globalité. La résistance doit partir de et développer les multiplicités, mais en aucun cas selon une direction ou une structure qui globalise, qui centralise les luttes. Un réseau de résistance qui respecte la multiplicité est un cercle qui possède, paradoxalement, son centre dans toutes les parties. Nous pouvons rapprocher cela de la définition du rhizome de Gilles Deleuze : 'Dans un rhizome on entre par n'importe quel côté, chaque point se connecte avec n'importe quel autre, il est composé de directions mobiles, sans dehors ni fin, seulement un milieu, par où il croît et déborde, sans jamais relever d'une unité ou en dériver ; sans sujet ni objet.'

'La nouvelle radicalité, ou le contre-pouvoir, ce sont bien sûr des associations, des sigles comme ATTAC, comme Act Up, comme le DAL. Mais ce sont surtout – et avant tout – une subjectivité et des modes de vie différents. Il y a des jeunes qui vivent dans des squats – et c'est une minorité de jeunes – mais il y a plein de jeunes qui pratiquent des solidarités dans leurs vies, qui n'ordonnent pas du tout leur vie en fonction de l'argent. Cela, c'est la nouvelle radicalité, c'est cette émergence d'une sociabilité nouvelle qui, tantôt, a des modes d'organisation plus ou moins classiques, tantôt non. Je pense qu'en France, ça s'est développé très fortement. Le niveau d'engagement existentiel des gens est énorme. (Benasayag, 2002 – my emphasis)

This is clearly a description of a new existential positioning, a radical refusal of power-based relationships and a clear departure from the old oppositional politics, where the protesters were using the same authoritarian principles in their midst, than those of the forces they were denouncing. Here are some further quotes, which highlight the new 'radical subjectivities':

Contrairement aux militants classiques, je pense que les choses qui existent ont une raison d'être, aussi moches soient elles...

Rien n'existe par accident et tout à coup, nous, malins comme nous sommes, nous nous disons qu'il n'y a vraiment qu'à décider de changer. Les militants n'aiment pas cette difficulté; ils aiment se fâcher avec le monde et attendre ce qui va le changer.

C'est toujours très surprenant: la plupart des gens ont un tas d'informations sur leurs vies, mais 'savoir', ça veut dire, en termes philosophiques, 'connaître par les causes', et donc pouvoir modifier le cours des choses.

Oui, l'anti-utilitarisme est fondamental. Parce que la vie ne sert à rien. Parce qu'aimer ne sert à rien, parce que rien ne sert à rien.

On voit bien cette militance un peu feignante qui se définit 'contre': on est gentil parce qu'on est contre. Non! ça ne suffit pas d'être contre les méchants pour être gentil. Après tout, Staline était contre Hitler! (Benasayag, 2002)

Conclusion

What have we tried to do in this chapter? Starting from the four types of sociality described by Alan Page Fiske, we have tried to show how peer-to-peer is a template for human relationships, that is expressed in a wide variety of fields of human endeavor (in the four quadrants of Ken Wilber), which mutually reinforce themselves. Peer-to-peer technology is the basic infrastructure of cognitive capitalism; it is a third mode of production not based on either profit or hierarchy; it is a new mode of distribution such as in the file-sharing networks; it is a new mode of organizing and conceiving cooperative relationships, expressed in a wide variety of social and political movements; it is a new way of feeling and thinking about the world. We have seen how peer-to-peer is not only a descriptive tool, but also a normative tool, which includes a critique of earlier modes of functioning, and a set of demands for new practices, such as for example in the field of spiritual experiencing. We have purported to show that peer-to-peer is therefore inextricably linked to both a potential re-enforcement of 'Equality Matching', and of a new domain of 'Communal Shareholding'.

If this chapter were to be continued, we would also have argued the following.

There is an increasing contradiction between the economic logic of cognitive capitalism, and its 'Market Pricing' dominance, and the social logic of new forms of cooperation, as well as the embeddedness of innovation in a general system of widespread public intelligence (the 'general intellect'). This creates a whole series of new conflict zones, new enclosures and disenclosures, struggles around the new public domain of knowledge, and about the very infrastructure of the hitherto peer-to-peer Internet. There are three potential scenarios of co-existence between the new cooperative sphere and the for-profit sphere: peaceful co-existence, information feudalism, and a new type of P2P society.

As we are not technological determinists, we are not saying that peer-to-peer technology will cause and determine the changes towards some utopian end state, but we do maintain that the technology both embeds, and reflects, a change in human mentality, and that it enables and empowers such changes, provided they are taken up by social movements. Furthermore, we believe that P2P, because it is such an essential part of the lives and practices of the new generations, is a powerful new discourse to reinforce or renew the emancipatory project of more equality and justice in the human lifeworld, adapted to the realities and forms of consciousness prevalent in cognitive capitalism. We also belief it can be usefully connected to older forms of 'Equality Matching' and 'Communal Shareholding', as defended by tribal movements defending their bio-agricultural inheritance and communal lands, by the labor movement, and by others, showing them that their demands, far from being only holdovers of an earlier era, are also pointers to a future where 'Market Pricing' and 'Authority Matching' are again balanced in a more equitable manner with the competing socialities of 'Equality Matching' and 'Communal Shareholding'.

References

Abbate, J. 1999. *Inventing the Internet*. Cambridge, MA: MIT Press.

Barbrook, Richard, Schultz, Pit. 2002. *The Digital Artisans Manifesto*, Hypermedia Research Center, downloaded 2002, http://www.hrc.wmin.ac.uk/hrc/theory/digitalartisans/t.1.1.html.

Benasayag, M.; Sztulwark, Diego. 2002. *Du contre-pouvoir*. Paris: La Decouverte.

Benasayag, Miguel with Lemahieu, Thomas. 2002. 'Resister "malgre tout"', *Peripheries*, downloaded 2002, http://www.peripheries.net/g-bensg.htm.

Cave, Damien. 2002. 'File sharing: Innocent until proven guilty', Salon, 13 June 2002, downloaded 2002, http://www.salon.com/tech/feature/2002/06/13/liebowitz/index.html.

Courant Alternatif. 2002. 'Argentine: entretien avec Miguel Benasayag', downloaded 2002, http://oclibertaire.free.fr/ca117-f.html.

Deleuze, G., Guattari, Felix. 1980. *Capitalisme et Schizophrenie*. Tome 2: Milles Plateaux. Paris: Ed. De Minuit.

Ferrer, J. 2001. *Revisioning Transpersonal Theory: A Participatory Vision of Human Spirituality*. Albany, NY: SUNY Press.

Alsop, Stewart. 2002. 'I want my file-served television', *Fortune*, 11 June 2002, downloaded 2002, http://www.fortune.com/fortune/alsop/0,15704,370066,00.html.

Free Software Association. 2004. 'The Free Software Definition', downloaded 2004, http://www.fsf.org/.

Hardt, M. 2002. 'Porto Alegre: Today's Bandung', *New Left Review* 14, downloaded 2002, http://www.newleftreview.net/NLR24806.shtml.

Hardt, M., Negri, Toni. 2001. *Empire*. Cambridge MA: Harvard University Press.

Heron, J.1998. *Sacred Science*. Ross-on-Wye: PCCS Books.

Himanen, P. 2002. *The Hacker Ethic and the Spirit of the Information Age*. New York: Random House.

Kramer, J., Alstad, Diane. 1993. *The Guru Papers: Masks of Authoritarian Power*. Berkeley, CA: Frog.

Lessig, L. 2004. Free Culture. *How Big Media uses technology and the law to lock down culture and control creativity*. New York: The Penguin Press.

Page Fiske, A. 1993. *Structures of Social Life*. New York: Free Press.

Paulson, Daryl. 2002. 'Daryl Paulson on Jorge Ferrer', Ken Wilber Online, downloaded 2002, http://wilber.shambhala.com/html/watch/ferrer/index.cfm/xid,76105/yid,55463210.

Raymond, E. 2001. *The Cathedral and the Bazaar. Musings on Linux and Open Source by an Accidental Revolutionary*. Sebastopol, CA: O'Reilly.

Ross, A. 2001. *No-Collar. The Humane Workplace and its Hidden Cost*. New York, NY: Basic Books.

Southern Poverty Law Center. 2001. 'The New Romantics', Intelligence Report, 101, downloaded 2002, http://www.splcenter.org/intel/intelreport/article.jsp?aid=236.

Stallman, R. 2002. *Free Software, Free Society: Selected Essays of Richard M. Stallman*. Boston, MA: Free Software Foundation.

The Economist. 2002. 'Watch this airspace', June 20, 2002 downloaded 2002, http://www.economist.com/printedition/displayStory.cfm?Story_ID=1176136.

Wallerstein, I. 2002. 'Porto Alegre, 2002', Commentary No. 82, Feb. 1, Fernand Braudel Center, downloaded 2002, http://fbc.binghamton.edu/82en.htm.

Wallerstein, I. 2004. *The Essential Wallerstein*. New York: New Press.

Weber, M. 2001.*The Protestant Ethic and the Spirit of Capitalism*. London: Routledge.

Weber, S. 2004. *The Success of Open Source*. Cambridge MA: Harvard University Press.

Wilber, K. 2001. *A Theory of Everything*. Boston: Shambhala.

8: From Virtual to Everyday Life

PAUL VERSCHUEREN

Of all the promises and prognoses made about old and new media, perhaps the most compelling has been the possibility of regenerating community through mediated forms of communication. (Jankowski, 2002: 34)

Introduction

About a decade ago, Howard Rheingold (1993) used the term 'virtual community' to bring the social aspects of computer-mediated communication under attention. He argued: *'whenever CMC technology becomes available to people anywhere, they inevitably build virtual communities with it, just as micro-organisms inevitably create colonies'* (1993: 6). Rheingold defined virtual communities as *'social aggregations that emerge from the Net when enough people carry on those public discussions long enough, with sufficient human feeling to form webs of personal relationships in cyberspace'* (1993: 5). His book told the history of a particular online community, the WELL (Whole Earth 'Lectronic Link), and showed how computers were not simply used to transmit information but to ritually connect people. He stressed that online social interactions were not simply based on self-interest but motivated by a desire for commonality.

The online community literature since Rheingold's book can be divided into three major types: the *utopian and dystopian discourses* from the early 1990s onwards, the *electronic field studies* from the mid-1990s onwards, and the *contextualized approaches* from the late 1990s onwards. These three types will be discussed here, roughly covering 10 years of research into online associations. The focus is on the virtual community as an analytical concept. It should be noted that many researchers have avoided the concept from the beginning. On the whole, however, the notion has had a powerful influence in academic as well as popular discourse.

Before Rheingold, research had focused on the differences between face-to-face communication and computer-mediated communication, and it had generally stressed the limitations of the latter. Rheingold's *The Virtual Community* moved researchers away from that perception, and also beyond the political and economic analyses of the 'Information Society' that were made in the 1990s (Robins & Webster, 1999). However, the concept also emphasized the distinction between newer online realities on the one hand and older offline realities on the other, associating the former with the global and the latter with the local. As I will later show, this had a narrowing effect on online community research.

Proulx and Latzko-Toth (2000: 7) see the concept of the virtual community as *a synthesis between, on the one hand, the growing fascination with the very word virtuality – as much on the popular imagination of engineers as on the imaginations of 'gurus' like Timothy Leary – and on the other hand, the term online community.*

According to Proulx and Latzko-Toth, the latter was introduced at the end of the 1960s by Licklider and Taylor (1968). It was only in the 1990s, however, that online associations became an important research topic in various disciplines, from psychology to philosophy.

An early and influential collection that focused on the new online associations was *Cybersociety* (Jones, 1995), later followed by *Cybersociety 2.0* (1998). The first edition of *Cybersociety* discussed such topics as social conduct, censorship and moderation on Usenet, and anonymity and identity construction through textual interaction. Baym's contribution, *The Emergence of Community in Computer-Mediated Communication* (1995), and Reid's *Virtual Worlds: Culture and Imagination* (1995), remain relevant introductions to the subject of online community formation.

Polemical Beginnings

Around the time of *Cybersociety*, the concept of the virtual community was turned into a buzzword. Many businesses began to use it as a model to generate profits. They gradually started to build 'community functions' into their websites. In popular discourse, almost every electronic system that provided one-to-one communication became a community. Virtual communities were said to exist within online conferences, list server groups, MUDs, MOOs,[1] and other interactive computer systems. These systems were heralded as liberating forms of communal experience, free from the constraints of physical reality and the physical body.

Critical voices reacted against this view, suggesting that computer-mediated communication merely offered a simulation of community or stimulated the development of narrow specialized interest groups. The critics claimed that virtual communities would contribute to isolation, to a decrease of human interdependence, to the decline of local communities in the physical world, and to the commodification of social behaviour (Boal, 1995; Kroker & Weinstein, 1994; Slouka, 1995; Stoll, 1995; Sardar, 2000).

These reactions are not surprising. Western discourses traditionally attach great significance to technological changes, either negatively or positively (e.g. Achterhuis, 1998). Technologies are received in a dystopian way, as a threat to contemporary ways of being, or praised in a utopian way as a liberating force. These extreme views surfaced frequently in discussions of virtual communities in the early 1990s. Utopists described earlier forms of community as too restrictive and welcomed the annihilation of time and space barriers. For them, the Internet offered more freedom, more equality and more prosperity (Benedikt, 1991; Gore, 1991; Negroponte, 1995; Stone, 1995; Turkle, 1995). Another utopian assumption was that civil society in virtual space would reclaim powers held by the state in geographical space (Barlow, 1996). Some utopists presented the Internet as a unifying force that would produce a single global 'cyberspace culture'. Although

Rheingold did not believe in '*a single, monolithic, online subculture*' (1993: 3), he also wrote that:

> [t]he small virtual communities still exist, like yeast in a rapidly rising loaf, but increasingly they are part of an overarching culture, similar to the way the United States became an overarching culture after the telegraph and telephone linked the states. (1993: 10)

In popular discourse, 'netiquette' and emoticons were cited as examples of such an 'overarching culture' although different netiquettes and emoticon systems exist. Euro-American and Japanese emoticons differ typographically as well as in the ways in which they are written, read and interpreted (Aoki, 1994). Japanese emoticons can be linked to double-byte character encoding, the Japanese typographic tradition, the Japanese *manga* (comic strips), Japanese body language, and other aspects of Japanese culture (Hiroe, 1999–2001; Aoki, 1994). The Japanese generally attach great value to politeness and appropriateness, and this is reflected in the Japanese emoticon system. It contains at least three different expressions of apology[2] for inappropriate behaviour, while the Euro-American system has not a single equivalent. The early utopian rhetoric of cyberspace suppressed these cultural differences by postulating a global culture with properties of its own. It separated the user from his or her locale, and presented this separation as liberation.

Dystopian critics rejected the idea of techno-liberation. They feared a decline of community and attached more value to local *Gemeinschaft*-like (Tönnies, 1979/1887) communities than to the newer online associations. They argued that people in geographical neighbourhoods are forced to live together, while members of global virtual communities can log on and log off whenever they want. According to the critics, the latter is problematic since it does not promote the responsibility, commitment and concern that geographical communities require. Their accounts were often inspired by science-fiction work such as *Neuromancer* (Gibson, 1984), which introduced the term 'cyberspace', and *Snow Crash* (Stephenson, 1992) with its own version of 'cyberspace' called 'Metaverse'. Following the publication of the anthology *Mirrorshades* (Sterling, 1988), these works became known as cyberpunk.[3] Cyberpunk fiction presents a world in which networked computers dominate everyday life. The focus is usually on underground cultures and struggles of alienated individuals against corporate powers. These popular representations stressed the alienating and dehumanising effects of computing technology, a theme inherited from earlier Western fiction (Huxley's *Brave New World*, Burgess' *A Clockwork Orange*, Orwell's *1984*, and so on).

Community and Identity
Dystopian critics claim that Internet technologies erode existing geographical communities. Utopian voices agree that communities are in decline but suggest that technologies can help to restore a sense of community (see, for instance, Rheingold, 1993). The idea of a community in decline, however, is a culturally specific and ideological construction. Social histories show that communities of the past were probably never as close-knit and cohesive as people sometimes like

to think (Laslett, 1999). Studies of nationalism (Anderson, 1983) and trans-nationalism (Hannerz, 1996) further indicate that face-to-face communication is less central to the development of communities than proponents of *Gemeinschaft*-like communities often claim. Indeed, many offline communities could be labelled 'virtual' since they are based on mediation and imagination. For instance, Stone (1995) calls the international academic community and the televisual community virtual too.

According to Thomas Bender (1982) the idea of a lost community recurs in different studies from the seventeenth century onwards. For Nancy, the idea has dominated Western thinking from Plato's *Republic* to Tönnies *Gemeinschaft und Gesellschaft* and beyond. Nancy calls it *'the most ancient myth in the Western world'* (1991: 10). In *The Inoperative Community*, he argues that the desire for an 'original' community is characteristic of Western discourses. These use the disappearance of community to explain the problems of contemporary life. According to Nancy (1991: 9):

> the lost, or broken, community can be exemplified in all kinds of ways, by all kinds of paradigms: the national family, the Athenian city, the Roman republic, the first Christian community, corporations, communes, or brotherhoods – always it is a matter of a lost age in which community was woven of tight, harmonious, and infrangible bonds.

Although the idea of a lost community frequently recurs, most researchers now accept that community is an ongoing process and that the disappearance of older community forms is accompanied by the emergence of newer kinds. Barry Wellman (e.g. 2001), for instance, suggests that community life has become privatized. Community is no longer established by going to public spaces but through person-to-person connectivity. Technologies, such as the telephone and e-mail, are used to establish and sustain these personalized networks. The concept of the 'personalized network' may avoid many problems associated with the traditional concept of community. Communities are often seen as isolated and bounded entities, but anthropologists dismiss such a view because it *'usually masks significant interactions between the individuals of that community and others, as well as the heterogeneity of the community itself'* (Wilson & Peterson, 2002: 455, referring to Appadurai, 1991). Online community studies often tend to focus on the ideational aspects of community only. The interpretive tradition[4] and the work of Benedict Anderson (1991/1983) in particular have stressed these aspects. However, as Amit (2002) notes, the ideational aspects should not be dissociated from actual social relations and everyday performances, something that was often the case in early discussions.

Since face-to-face communication differs across cultures, we may expect to find cultural differences in e-mailing, MUDding, chatting, and other forms of electronic association. These differences, however, were usually not discussed in the utopian and dystopian discourses of the early 1990s. Both tended to treat the Internet as a single, totalising force and paid little attention to the differences between the various Internet technologies. For instance, the Internet was said to promote 'identity play' in virtual communities. This was heralded as liberation by many utopists, and dismissed as a simulation of the self by dystopists. However, Goffman's work (1987/1959) suggests that identity play is not characteristic of online behaviour, but a general feature of social life. The differences between offline and online behaviour therefore appear to be of degree rather than of kind. Furthermore, identity performance in e-mail exchange is quite different from identity performance in MUDs or MOOs. In regular e-mail, identities tend to be more or less fixed. The WELL allows multiple representations of self, but these have to be related to a single, fixed user-ID (Rheingold, 1993). MUDs and MOOs are usually oriented towards fantasy and play, and allow for experimentation. These electronic environments have a liminal quality (Turner, 1970), allowing participants to explore roles and activities that are normally impossible or socially unacceptable. As in other liminal circumstances, such as traditional carnivals, identity play and 'gender swapping' are to be expected here.

The Real/Virtual Dichotomy

Utopian and dystopian discourses assume that social effects flow naturally from the technology employed. This deterministic vision presupposes that technologies can shape social and cultural worlds from scratch. But something has always gone on before. Users inevitably carry with them a particular history, education, gender, class, ethnic background, and so on. Even liminal, role-playing experiences relate to a previous socio-cultural state (Turner, 1970). Thus, social behaviour, norms and values cannot be abstracted from their local, historical and socio-cultural context, as quite a few of the earlier studies seemed to suggest. Agre (1999: 4) argues that

> so long as we focus on the limited areas of the Internet where people engage in fantasy play, we miss how social and professional identities are continuous across several media, and how people use those several media to develop their identities in ways that carry over to other settings.

Utopian and dystopian discourses presuppose a too sharp distinction between electronic and face-to-face realities. Proulx and Latzko-Toth (2000) call the latter a *'discourse of denigration'* because it subordinates the 'virtual' to the 'real'. The former is its reversal since it *'sees virtuality as the 'resolution' of a world overwrought by imperfection as the consequence of its presence, which is but a subset of the universe of possibilities – and therefore an unavoidable impoverishment'* (Proulx and Latzko-Toth, 2000: 5). Both discourses fail to see how pre-existing socio-cultural contexts are inextricably intertwined with Internet technologies. Wilson and Peterson (2002: 456) observe that

> [a]n online/offline conceptual dichotomy [for example Castells' (1996) 'network society'] is also counter to the direction taken within recent anthropology, which acknowledges the multiple identities and negotiated roles individuals have within different socio-political and cultural contexts.

Social shaping of technology studies (e.g. Bijker, Hughes & Pinch, 1987; Latour, 1996; 1999; Law & Hassard, 1999; MacKenzie & Wacjman, 1985) indicate that the usage and the development of technologies is related to socio-cultural contexts. Rejecting simple causalities, these studies recognize technologies as agents of change, but also point out that technological effects are strongly dependent on the socio-cultural context in which the technologies are used and have been developed.

One social shaping approach, Actor Network Theory (ANT), is premised on the idea that technological, symbolical, and corporeal spaces are co-constructing each other. These constitute a connected space, a complex eco-social system, in which the meaning of an entity depends on its relationship with other entities. ANT proposes a generalized form of material-semiotics, derived from the work of Saussure and Greimas. Objects and subjects are considered as actants (which are simultaneously networks), constructed by each other and linked together in a single, connected space. Each actant is constituted through a web of influences and connections. To study an actant is to describe how the actant relates to other actants (other users, other humans, technologies, localities, and so on), and how powerful actants define and control a network through their various relations. Typically, ANT rejects a human-centred approach and treats all actants – both human and non-human – in a methodologically neutral way. Seen from this perspective, virtual communities are, and consist of, actants within a much broader network or context than their association with the bounded world of 'cyberspace' suggests.

Online Ethnographies

Most of the electronic field studies of the 1990s did not take this contextual view. Contrary to earlier utopian and dystopian accounts, they looked closely at the social interactions inside virtual communities, in an empirical way, and covering a great variety of environments, ranging from health and religious communities to digital cities. Many field studies tried to find out how, and to what extent, these interactions create a sense of community. Textual conversations of e-mail lists, MOOs and MUDs were downloaded by the researcher, sometimes interviews with community members were added, and the data was subsequently analysed in a quantitative or qualitative fashion.

One example of an in-depth ethnography and textual analysis is Lynn Cherny's *Conversation and Community: Chat in a Virtual World* (1999). Her work gives a detailed analysis of how a sense of community was created, maintained and reproduced in ElseMOO. Cherny, who was a member of ElseMOO before she started her study, used participant observation, a survey, and conversation analysis to investigate her online environment. The major part of her book is devoted to conversation and the formation of social relations. It focuses on general aspects of communication, such as register and turn taking, but also deals with medium specific issues such as 'emoting'.[5] Cherny's study, unlike utopian and dystopian narratives, draws a subtle picture of life inside an online community. It shows that online communities resemble geographical communities in multiple ways. Both develop a sense of belonging by establishing common values and beliefs, a common rhetoric, identity and ideology, a (mythical) history, social hierarchies, boundary mechanisms, and so on. Her work confirms the utopian claim that online communities can be more than narrowly defined interest groups. Several aspects of traditional place-based communities can be found in virtual communities, including ongoing interaction and reciprocity, common rituals, rules, and norms, social memory (for instance, histories told in FAQ lists), chance meetings, a sense of local space, identity and boundary politics, conflict resolution, and so on. Cherny's work focuses on social stratification, power distribution, and the establishment of authority and popularity. Cherny found important differences between ordinary and more powerful community members. The latter had an excellent command of ElseMOO's own register and contributed to the community in important ways, for instance as administrators or 'wizards'.

In *Life on the Screen*, Sherry Turkle (1995) discusses how online environments allow for experimentation with the self, and these experiments are seen as potentially liberating. Contrary to her utopian vision of techno-liberation, electronic field research has pointed out various forms of racism, gender and other kinds of discrimination in virtual communities. There is now a considerable body of literature showing that the Internet does not remove individuals from cultural differentiation and existing power structures (Ebo 1998; Escobar, 1996). Feminist scholars such as Wise (1997) discussed gender discriminations and the reproduction of patriarchal forms of oppression. Burkhalter (1999) showed that participants in Usenet groups often want to be known by their 'racial' identity. His conclusion was that *'racial stereotypes may be more influential and resilient on the Usenet'* (Burkhalter, 1999: 74). According to Nakamura (2000), many MUD characters are based on racial stereotypes. She has indicated forms of orientalism, and has pointed out ways in which stereotypical user identities are inscribed into interface designs. For instance, MOO characters may be 'white' by default, making all the others accountable for their 'non-white' identity. A number of these critical approaches were collected in *Communities in Cyberspace* (Smith & Kollock, 1999).

Bodies remain important in online communities, even though they may be re-imagined. One of the common questions in chatting environments, 'asl?' (*a*ge, *s*ex, *l*ocation?), illustrates the point that the local and corporeal do matter in virtual environments. Many electronic field studies tend to neglect this pre-given corporeality and assume that the Internet allows for entirely disembodied ways of being. They focus on conversations without a deep understanding of the participants' everyday life situation, and without any certainty about the participants' demographic profile. The focus is predominantly on intra-community behaviour, while the inter- and extra-community dimensions remain absent or underexposed. Analyses of online conversations do not tell much about the ways in which individuals move between communities. Neither do they reveal much about the ways in which this behaviour is embedded in historical and socio-cultural contexts. What has usually been left out is:

> the link between historically constituted socio-cultural practices within and outside of mediated communication and the language practices, social interactions, and ideologies of technology that emerge from new information and communication technologies.
> (Wilson & Peterson, 2002: 453)

Until the late 1990s, most ethnographies took an *'Internet as culture'* (Hine, 2000) perspective. They dealt with the symbolic construction of online community and treated the Internet as a context for social relations. This moved research away from an instrumentalist perspective. However, the Internet as a cultural artefact (Hine, 2000), rooted in cultural and historical conditions, remains largely unexplored. The view of technology as context rather than as a cultural artefact is characteristic of much anthropological work (Pfaffenberger, 1992; Wilson & Peterson, 2002). Only recently, anthropologists have begun to explore the cultural dimensions of media technologies.

The construction of the Internet as an artefact relates to the 'social-shaping-of-technology' perspective mentioned earlier. Bruno Latour (e.g. 1992, 1996) has argued that the values, beliefs, norms, goals, social attitudes and practices of dominant social groups enter into technological artefacts. Consequently, artefacts discipline: they are likely to reinforce the cultural and social aspects that have been entered. Latour calls this prescription. For example, personal computer interfaces reflect the world of office workers, with an emphasis on bureaucratic tasks such as filing. The default 'white' identity in virtual environments mentioned earlier is another example of prescription. As this example shows, prescriptions are the politics of an artefact: they tell us what users should do and look like, what the moral codes of the community are, and they define who is inside and who is outside the community. Prescription is never absolute: disciplining does not happen in any deterministic way since prescriptions need not be subscribed to. They can be contested, resisted, and de-inscribed. It is obvious that these politics of prescription, contestation, and de-inscription cannot be explained adequately from interactions within the online community alone. They need to be related to wider contexts.

Everyday Life Approaches

In the second half of the 1990s, researchers began to contextualize Internet technologies more thoroughly. While community network studies had obviously focused on linkages between online and offline realities, some researchers now began to explore these links in other cases too (e.g. Wellman & Haythornthwaite, 2002; Miller & Slater, 2000). This marked a shift from cyberspace to everyday realities. This coincided with the *normalization of cyberspace*' (Margolis & Resnick, 2000) in larger parts of the Western world as a result of longer and more frequent usage of the Internet and the convergence of information and communication technologies, which is oriented towards integration into everyday life.

Researchers who take an everyday life perspective study community ties regardless of their locality, and all the technologies (telephone, Internet, and so on) used to establish them. This seems more productive than the one community/one technology approach of earlier online ethnographic studies. Wilson and Peterson note that '*an anthropological approach is well suited to investigate the continuum of communities, identities, and networks that exist*' (2002: 456, my emphasis). The concept of the *'personalized community*', proposed by the sociologist Wellman (e.g. Wellman 1997; Wellman & Haythornthwaite, 2002), seems well suited for the exploration of this continuum. This concept indicates that individuals in contemporary Western societies do not live in all-encompassing communities in public spaces but spend their lives mostly in networks established in private spaces. A personalized network consists of relations with kin, friends, neighbours, and organizations, and includes memberships in multiple and partial communities. Internet technologies are means, among many others, to establish and maintain these relations. But they are also more than that. A consideration of technologies as actants avoids their reduction to mere tools as well as to mere contexts. Network analyses, in Wellman's (1997: 179) view:

> trace the social relationships of those they are studying, wherever these relationships go and whomever they are with. Only then do network analysts look to see if such relationships actually cross formal group boundaries. In this way formal boundaries become important analytic variables rather than a priori analytic constraints.

The turn from studies of virtual communities as bounded units towards a focus on the integration of computer-mediated communication into everyday life contexts carries with it obvious and substantial methodological benefits. For instance, demographic reliability increases. In online textual environments, identities are difficult to verify. Various avatars (characters) may represent a single individual at various times in the same online community. These avatars are hard to link and track from an online perspective, but they can easily be associated with each other when the physical individual is taken as the starting point.

An everyday life perspective also helps to critically examine the common accusation that Internet technologies, by eradicating time and space boundaries, separate individuals from their face-to-face relationships and communities. Contrary to dystopian assertions that virtual communities may be detrimental to the strength of geographical communities, Hampton and Wellman's (2002) everyday life study of a Toronto suburb called Netville states that Internet technologies reinforce existing place-based communities. A recent study by Matei and Ball-Rokeach (2002: 420) further holds that *'a higher level of belonging to real communities translates into a higher propensity for interaction online'*.[6] According to this study, individuals are more likely to make online friends when they know more people in the neighbourhood and believe that they live in an area characterized by neighbourliness.

A focus on everyday life contexts may reveal new social patterns, and will move digital divide discussions beyond matters of access towards a consideration of the integration of communication tools into daily life. For instance, Howard, Rainie and Jones (2002) show that of those with Internet access, more of the men, whites, higher-educated, higher-income earners, and more experienced users are effectively online on any given day.

The studies mentioned above, collected in *The Internet in Everyday Life* (Wellman & Haythornthwaite, 2002), confirm that online interactions are not a substitute for offline relationships but tend to extend the latter and increase interaction between people. Maria Bakardjieva has used the term *'immobile socialization'* to describe the use of the Internet in this *'process of collective deliberation and action in which people engage from their private realm'* (2003: 291). She uses the term 'immobile socialization' to contrast the Internet with broadcast technology and the automobile that stimulated the withdrawal of the middle class from public spaces, a process described by Williams (1974) as *'mobile privatization'*.

The findings, mainly based on quantitative data, provide a broad overview, but many details of personal lives, practices and experiences are kept out of sight. Thick ethnographic descriptions of how people build and perform social networks in everyday life – with and without the aid of Internet technologies – are missing in this volume. An example of such a thick ethnography is Daniel Miller and Don Slater's *The Internet: An Ethnographic Approach* (2000). Their study focuses on Internet usage in Trinidad and deals with a wide variety of issues, including kinship, national identity, business, politics and religion. The authors reject the assumption that the virtual is disembedding the corporeal. By contrast, they start from the premise that Internet is a collection of '*numerous new technologies, used by diverse people, in diverse real world locations*' (2000: 1), and consequently take into consideration the specific nature of Trinidadian culture as well as the diversity of the technology. Miller and Slater's work is a study of personal communities or networks in which the distinction between online and offline worlds are blurred. Trinidadian social associations, both online and offline, are sustained by multiple means, including the Internet. The authors stress that culture influences the ways in which individuals relate to technologies. For instance, they argue that ICQ fits Trinidadian culture particularly well, because it relates to the anti-structural offline habits of hanging around without a specific purpose, known in Trinidadian culture as 'liming'. In a number of other ways too, Miller and Slater show how relationships on the Internet are closely linked to more traditional forms of association. Contrary to dystopian claims, Miller and Slater argue that the Internet is strengthening private communities, such as the nuclear and extended family in Trinidad. The authors also explain how Trinidadians re-imagine their offline locality on the Internet. Their work demonstrates that Internet technologies are used to reconnect people to places rather than 'liberate' them from their geographical localities.

Rather than innovating a new kind of ethnography, Miller and Slater treated the Internet as part of Trinidadian material culture. Their qualitative work is premised on the same ideas as *The Internet in Everyday Life*: Internet technologies are embedded in local everyday contexts, and do not produce separate, isolated 'virtual communities'.

Conclusion

In the first half of the 1990s, the concept of the virtual community broadened the view on information and communication technologies. It shifted attention from the technological, communicative, political and economic aspects of computer networks towards the social and cultural ones. The concept of the virtual community, however, also separated the Internet from local everyday life contexts. It stressed the Internet *as a global context for social relations* rather than *a medium used within particular local contexts*. Several metaphors, such as 'the information superhighway' and 'cyberspace', contributed to the perception of the Internet as a separate sphere, and to its mythologization as a world of better social relations, more prosperity, and more freedom. It is probably no coincidence that this discourse of an alternative space emerged at the end of the colonial era and at a time of great uncertainty about the world's ecological system (Escobar, 1996; Sardar, 2000; Gunkel & Gunkel, 1997). Cyberspace and the virtual community can be seen as the Western middle-class response to these historical circumstances. In the early 1990s, cyberspace and the virtual community created the illusion of a better, entirely controllable, anthropocentric, and a-historical world.

Early utopian and dystopian discourses treated the Internet as an outside force that would shape new, virtual communities. Virtual communities, however, do not flow naturally from the technology employed. Their characteristics cannot be derived in any straightforward way from the possibilities and constraints offered by the technology. Social shaping of technology studies suggests that media technologies are the result of social choices. Using and developing these technologies is a culturally specific process, located in historical and social contexts, although dominant ideologies about new technologies frequently suggest the opposite. Consequently, the online/technological/global and the offline/corporeal/local should be treated as a single, connected, heterogeneous space. A material-semiotic approach, as proposed by ANT, can help us to better understand this connected space and the heterogeneous nature of its entities. ANT focuses on the ongoing process of interaction between technology and society: the Internet, its usage and development as the result of socio-cultural contexts. This aspect has seldom been explored in ethnographies of virtual communities, and the interaction between the 'Internet as culture' and the 'Internet as a cultural artefact' has received even less attention.

The concept of virtual community reduced Internet ethnography in the 1990s to the study of the 'Internet as culture'. The focus was almost entirely on social behaviour in bounded online spaces. Electronic field studies demonstrated how community cultures emerged from online interactions, but they usually did not show how these communities were related to broader social, cultural and political contexts. Since they conceptualized 'community' as something that can be spatially demarcated, they resembled traditional neighbourhood community studies much more than their 'exotic' topics at first sight suggested. Both fixed community and community members in a particular (electronic or geographical) space. This perspective does not reflect the way people incorporate technologies in their daily life, and the perception that people generally do not live in bounded communities.

The alternative, everyday life perspective that is gaining prominence assumes that social behaviour is embedded in wider networks, and that these networks are sustained by various technologies and social practices. This view stresses that the Internet continues, maintains and extends relationships, that it is used to perform one's identity (Goffman, 1959) and to spin webs of significance (Geertz, 1973) in old as well as new ways. People will continue to meet in online environments, but these are not entirely separate from their physical lives and corporeal contexts. The socialization into online communities, the negotiation, reproduction and contestation of identities, and the integration of computing technologies into everyday practices are some of the issues that cannot be understood as long as the online/offline dichotomy is sustained (Wilson & Peterson, 2002). The anthropological work of Miller and Slater and the work of the sociologist Wellman indicate ways in which these issues can be adequately dealt with.

Notes

1   In its original form, a MUD (Multi-User Dimension or Dungeon) is a multi-user fantasy game that simulates the physical world by means of textual descriptions. MUD players interact with each other and with this simulated environment. A MOO (Multi-User Object Oriented) is a further development of a MUD. In MOOs, players can create objects, including characters. MUDs and MOOs are now used for gaming as well as social interaction. MUDs and MOOs have led to the development of commercial multiplayer online role-playing games (MMORPG).

2   These Japanese expressions are: (^o^;>) *excuse me!*, (_o_) *I'm sorry*, (*^_^*;) *sorry*. Japanese emoticons are not read 'horizontally' or sideways as Euro-American emoticons are. They are read in the ordinary 'vertical' position. Hiroe, 1999–2001.

3   The term is from Bruce Bethke's short story *Cyberpunk* (1980). Cyberpunk literature can be associated with a wide range of popular representations, including those from films such as *Blade Runner, Total Recall* or from Masamune Shirow's manga *Kôkaku kidôtai* (*Ghosts in the Shell*).

4   See Cohen, who describes communities as '*worlds of meaning in the* minds *of their members*' (1985: 20 – my emphasis).

5   Emoting refers to the descriptions of actions and moods as substitutes for the physical signals in face-to-face communication.

6   See Matei and Ball-Rokeach (2002) for a list of other studies that confirm these findings.

References

Achterhuis, H. 1998. *De erfenis van de utopie*. Amsterdam: Ambo.

Agre, P. 1999. 'Life after Cyberspace', *EASST Review* 18: 3–5.

Amit, V. 2002. 'Reconceptualizing Community', pp. 1–20 in V. Amit (ed.) *Realizing Community*. London and New York: Routledge.

Anderson, B. 1991/1983. *Imagined Communities: Reflections on the Origin and Spread of Nationalism*. London and New York: Verso.

Aoki, K. 1994. 'Virtual Communities in Japan', paper presented at the Pacific Telecommunications Council 1994 Conference. Downloaded on 10 January 2003 from http://www.ibiblio.org/pub/academic/communications/papers/Virtual-Communities-in-Japan.

Appadurai, A. 1991. Global Ethnoscapes: Notes and Queries for a Transnational Anthropology, pp. 191–210 in R.G. Fox (ed.) *Recapturing Anthropology: Working in the Present*. Santa Fe, NM: School of American Research Press.

Bakardjieva, M. 2003. 'Virtual Togetherness: An Everyday-Life Perspective', *Media, Culture & Society* 25: 291–313.

Barlow, J. P. 1996. 'A Declaration of Independence of Cyberspace'. Dowloaded on 10 January 2003 from http://www.eff.org/~barlow/Declaration-Final.html.

Baym, N. K. 1995. 'The Emergence of On-line Community', pp. 138–163 in S.G. Jones (ed.) *Cybersociety: Computer-Mediated Communication and Community*. Thousand Oaks, CA: Sage Publications.

Bender, T. 1982. *Community and Social Change in America*. Baltimore and London: The Johns Hopkins University Press.

Benedikt, M. 1991. 'Introduction', pp. 1–25 in M. Benedikt (ed.) *Cyberspace: First Steps*. Cambridge, MA: MIT Press.

Bethke, B. 1980. 'Cyberpunk'. Downloaded on 10 January 2003 from http://project.cyberpunk.ru/lib/cyberpunk/. First published in Amazing Science Fiction Stories 57(4), November 1983.

Bijker, W. E., Hughes, T. P., Pinch, T. J. 1987. *The Social Construction of Technological Systems: New Directions in the Sociology and History of Technology*. Cambridge, MA: MIT Press.

Boal, I. A. 1995. 'A Flow of Monsters: Luddism and Virtual Technologies', pp. 3–15 in J. Brook & I.A. Boal (eds.) *Resisting the Virtual Life: The Culture and Politics of Information*. San Francisco: City Lights.

Burkhalter, B. 1999. 'Reading Race Online: Discovering Racial Identity in Usenet Discussions', pp. 60–75 in M. A. Smith and P. Kollock (eds.) *Communities in Cyberspace*. London and New York: Routledge.

Cherny, L. 1999. *Conversation and Community. Chat in a Virtual World*. Stanford, CA: CSLI Publications.

Cohen, A. P. 1985. *The Symbolic Construction of Community*. London: Tavistock.

Cohill, A. M., Kavanaugh, A. L. 1997. *Community Networks: Lessons from Blacksburg, Virginia*. Boston: Artech House.

Ebo, B. L. 1998. *Cyberghetto or Cybertopia? Race, Class, and Gender on the Internet*. Westport, CT: Praeger.

Escobar, A. 1996. 'Welcome to Cyberia: Notes on the Anthropology of Cyberculture', pp. 111–137 in Z. Sardar & J. R. Ravetz (eds.) *Cyberfutures: Culture and Politics on the Information Superhighway*. New York: New York University Press.

Geertz, C. 1973. *The Interpretation of Cultures*. New York: Basic Books.

Gibson, W. 1984. *Neuromancer*. New York: Ace Books.

Goffman, E. 1987/1959. *The Presentation of Self in Everyday Life*. Harmondsworth: Penguin Books.

Gore, A. J. 1991. 'Information Superhighways. The Next Information Revolution', *The Futurist* 25: 21–23.

Gunkel, D. J., Gunkel, A. H. 1997. 'Virtual Geographies: The New Worlds of Cyberspace', *Critical Studies in Mass Communication* 14: 123–137.

Howard, P. E. N., Rainie, L., Jones, S. 2002. 'Days and Nights on the Internet', pp. 45–73 in B. Wellman & C. Haythornthwaite (eds.) *The Internet in Everyday Life*. Oxford: Blackwell Publishing.

Hine, C. 2000. *Virtual Ethnography*. London: Sage Publications.

Hiroe, T. 1999–2001. 'Japanese smileys' Downloaded on 10 January 2003 from http://club.pep.ne.jp/~hiroette/en/facemarks/index.html.

Hampton, K. N., Wellman, B. 2002. 'The Not So Global Village of Netville', pp. 345–372 in B. Wellman & C. Haythornthwaite (eds.) *The Internet in Everyday Life*. Oxford: Blackwell Publishing.

Hannerz, U. 1996. *Transnational Connections*. London: Routledge.

Jankowski, N. W. 2002. 'Creating Community with Media: History, Theories and Scientific Investigations', pp. 34–49 in L. A. Lievrouw & S. Livingstone (eds.) *Handbook of New Media. Social Shaping and Consequences of ICTS*. London, Thousand Oaks and New Delhi: Sage Publications.

Jones, S. G. (ed.) 1995. *Cybersociety: Computer-Mediated Communication and Community*. Thousand Oaks, CA: Sage Publications.

Kolko, B. E., Nakamura, L., Rodman, G. B. (eds.) 2000. *Race in Cyberspace*. New York: Routledge.

Kroker, A., Weinstein M. A. 1994. *Data Trash: The Theory of the Virtual Class*. New York: St. Martin's Press.

Laslett, P. 1999. *The World We Have Lost: Further Explored*. London: Routledge.

Latour, B.1992. 'Where are the missing masses? A sociology of a few mundane artifacts', pp. 225–258 in W. E. Bijker and J. Law (eds.) *Shaping Technology/Building Society*. Cambridge, MA: MIT Press.

Latour, B. 1996. *Aramis, or, The Love of Technology*. Cambridge, MA: Harvard University Press.

Law, J., Hassard, J. 1999. *Actor Network Theory and After*. Oxford: Blackwell Publishers/The Sociological Review.

Licklider, J. C. R., Taylor, R. W. 1968. 'The Computer as a Communication Device'. Downloaded on 10 January 2003 from ftp://ftp.digital.com/pub/DEC/SRC/research-reports/SRC-061.pdf. Originally published in *Science and Technology*, April 1968.

MacKenzie, D., Wajcman, J. 1985. *The Social Shaping of Technology*. Milton Keynes: Open University Press.

Margolis, M., Resnick, D. 2000. *Politics as Usual: The Cyberspace 'Revolution'*. London: Sage Publications.

Matei, S., Ball-Rokeach, S. J. 2002. 'Belonging in Geographic, Ethnic, and Internet Spaces', pp. 404–427 in B. Wellman & C. Haythornthwaite (eds.) *The Internet in Everyday Life*. Oxford: Blackwell Publishing.

Miller, D., Slater, D. 2000. *The Internet: An Ethnographic Approach*. Oxford: Berg.

Nakamura, L. 2000. 'Race in/for Cyberspace. Identity Tourism and Racial Passing on the Internet', pp. 712–720 in D. Bell & B. M. Kennedy (eds.) *The Cybercultures Reader*. London: Routledge.

Nancy, Jean-Luc. 1991. *The Inoperative Community*, P. Connor (ed.), transl. P. Connor, L. Garbus, M. Holland, S. Sawhney. Minneapolis and London: University of Minnesota Press.

Negroponte, N. 1995. *Being Digital*. London: Hodder and Stoughton.

Proulx, S., Latzko-Toth, G. 2000. 'The Curious Virtuality of the Virtual: On the Social Scientific Category of "Virtual Communities"'. Paper presented at the 1st Conference of the Association of Internet Researchers, Lawrence, Kansas.

Pfaffenberger, B. 1988. 'The Social Meaning of the Personal Computer: Or Why the Personal Computer Revolution was No Revolution', *Anthropological Quarterly* 61: 39–47.

Pfaffenberger, B. 1992. 'Social Anthropology of Technology', *Annual Review of Anthropology* 21: 491–516.

Reid, E. 1995. 'Virtual Worlds: Culture and Imagination', pp. 164–183 in S. G. Jones (ed.) *Cybersociety: Computer-Mediated Communication and Community*. Thousand Oaks, CA: Sage Publications.

Rheingold, H. 1993. *The Virtual Community. Homesteading on the Electronic Frontier*. Reading: Addison-Wesley.

Robins, K., Webster, F. 1999. *Times of the Technoculture: From the Information Society to the Virtual Life*. London: Routledge.

Sardar, Z. 2000. 'Alt.civilizations.faq. Cyberspace as the Darker Side of the West', pp. 732–752 in D. Bell & B. M. Kennedy (eds.) *The Cybercultures Reader*. London and New York: Routledge.

Slouka, M. 1995. *War of the Worlds: Cyberspace and the High-Tech Assault on Reality*. New York: Basic Books.

Smith, M. A., Kollock, P. 1999. *Communities in Cyberspace*. London and New York: Routledge.

Stephenson, N. 1992. *Snow Crash*. New York: Bantam Books.

Sterling, B. 1986. *Mirrorshades: the Cyberpunk Anthology*. New York: Arbor House.

Stoll, C. 1995. *Silicon Snake Oil: Second Thoughts on the Information Highway*. New York: Doubleday.

Stone, A. R. 1995. *The War of Desire and Technology at the Close of the Mechanical Age*. Cambridge, MA: MIT Press.

Tönnies, F. 1979/1887. *Gemeinschaft und Gesellschaft: Grundbegriffe der reinen Soziologie*. Darmstadt: Wissenschaftliche Buchgesellschaft.

Turkle, S. 1995. *Life on the Screen: Identity in the Age of the Internet*. New York: Simon & Schuster.

Turner, V. W. 1970. *The ritual process: structure and anti-structure*. Chicago: Aldine.

Wellman, B. 1997. 'An Electronic Group is Virtually a Social Network', pp. 179–205 in S. Kiesler (ed.) *Culture of the Internet*. Hillsdale, NJ: Lawrence Erlbaum.

Wellman, B., Haythornthwaite, C. (eds.) 2002. *The Internet in Everyday Life*. Oxford: Blackwell Publishers.

Williams, R. 1974. *Television: Technology and Cultural Form*. London: Fontana.

Wilson, S. M., Peterson, L. C. 2002. 'The Anthropology of Online Communities', *Annual Review of Anthropology*, 31: 449–467.

Wise, P. 1997. 'Always Already Virtual: Feminist Politics in Cyberspace', pp. 179–196 in D. Holmes (ed.) *Virtual Politics. Identity and Community in Cyberspace*. London, Thousand Oaks and New Delhi: Sage Publications.

9: Shifting from Equity to Efficiency Rationales: Global Benefits Resulting from a Digital Solidarity Fund

CLAUDIO FEIJÓO GONZÁLEZ, JOSÉ LUIS GÓMEZ
BARROSO, ANA GONZÁLEZ LAGUÍA, SERGIO RAMOS
VILLAVERDE, DAVID ROJO ALONSO

Introduction

Realising the potential of the Information Society requires an adequate infrastructure, a sine qua non condition for usage. One of the key principles of the WSIS *Declaration of Principles* is entitled '*Information and communication infrastructure: an essential foundation for an inclusive information society*'. Item 21 stated that '*Connectivity is a central enabling agent in building the Information Society. Universal, ubiquitous, equitable and affordable access to ICT infrastructure and services, constitutes one of the challenges of the Information Society and should be an objective of all stakeholders involved in building it.*'

Investment in broadband, which requires a significant improvement of the existing infrastructures or even a new network deployment, will mainly come from the private sector. The public sector must help create a favourable environment and stimulate demand. However, it is unlikely that operators will maintain any interest outside grouped-and-profitable-customer-filled urban areas. Isolated and rural areas may have to wait quite some time until they can enjoy, not the arrival of effective competition, but any broadband connection. So, governments must also take action on the supply-side of the market.

This is a problem faced by developed countries, since they need to avoid the extension of the digital divide which threatens leaving their remote or depressed regions behind. However, this problem is especially serious in less developed countries. When the national sector is incapable of meeting such needs, aids and loans become the primary, if not the only, solution. Without any foreign aid, the objective of achieving general access to telecommunication services seems to be quite far away, despite the existence of mechanisms guaranteeing that access, at least in theory, in almost all of them.

Programmes fostering a general economic development must allocate special importance to telecommunications. Nonetheless, more specific actions are required. As expected, this fact has been highlighted in a summit such as the WSIS that establishes as its first declaration the *'desire and commitment to build a people-centred, inclusive and development-oriented Information Society'*. The WSIS *Declaration of Principles* calls for digital solidarity, both at the national and international levels (Item 17). However, the section dealing with international cooperation represents no more than a simple declaration of intentions.

It is our idea that without the richest countries becoming aware of the advantages they would receive by supporting these actions, the chances of building a policy that proves actually effective are very small. The consideration of externalities and the provision of global public goods open a path that can transform a perspective of discretional and insufficient donations into a cooperation model based on self-interest for the global development of the networks.

The chapter is structured as follows: the role of governments in network development is analysed in section 2, giving special consideration to the scenario in less developed countries. Section 3 describes the reasons why the development of advanced telecommunication networks in all countries would generate global benefits. The results of this section lead us to proposing, in section 4, a new political orientation which would replace 'aid' with 'cooperation'.

Mechanisms Allowing Generalised Access to Telecommunication Services

From Monopoly-Based Public Service to Universal Service

The mechanisms that intended to guarantee generalized access to telecommunications have existed almost since the beginning of network deployments. One of the fundamental goals of regulated national monopolies was the provision of voice communications to all citizens at uniform (i.e. geographically averaged) 'affordable prices'. The network development plans were historically funded by cross-subsidies within the regulated price structure of national monopolies. Long-distance calls and customers in urban areas subsidized telephone access (and sometimes local calls) and customers in rural (high-cost, scarcely populated) areas. However, for a greater part of the century, the service extension commitment was in most countries more implicit than explicit, and interpreted from a basically voluntaristic perspective by most governments.

In recent decades, the dramatic technological progress as well as the changes observed in the regulatory framework have completely transformed the telecommunications sector. Competition forces (even when benefiting the telecommunications industry as a whole) undermine the sustainability of cross-subsidies and then destroy the traditional funding mechanism of network deployment.

At the time market opening was set out, it was necessary to elucidate whether, under these conditions, the competing industry, by itself and without regulations, could provide the service under reasonable conditions to all that requested it. Since it was predictable, as later confirmed, that competition processes would extend unevenly and would target the profitable segments of the market, finding a system that continued to guarantee access to essential services seemed necessary. This is no other than the universal service, which emerges, thus, as an attempt to reconcile the principles of public service with those of a market economy.

A unique global definition of universal service does not exist. What does exist, however, is an agreement on the fact that the basic core of the concept should usually cover the availability in the national scope of specific services for which non-discriminatory access and generalized economic affordability are guaranteed (ITU, 1998).

The approach to universal service is quite pragmatic. Despite a certain uniformity of the definitions included in most telecommunications legislations, the practical construction of universal service differs from one country or region to another, and even inside the same country when the context varies (ITU, 1994). This is nothing new: even before the figure of universal service in telecommunications appeared in its modern sense, the objectives of universality had changed through time according to technological development, infrastructure deployment levels and user requirement perception (Bardzki & Taylor, 1998). Furthermore, the WTO agreement on basic telecommunications services respects each country's faculty to define its own domestic universal service obligations and finance them in the way it considers most suitable.[1]

Universal access does not necessarily imply a line for every household. Establishing a shared or community access is the universal service modality chosen by many developing countries where the objective of one telephone per household is a Utopian plan (see Falch and Anyimadu, 2003). The requirement can be connected to a distance (one access point available in less than 'x' kilometres), 'trip time' or population size datum (refer to ITU, 1998). According to Item 23 of the WSIS *Declaration of Principles 'the establishment of ICT public access points in places such as post offices, schools, libraries and archives, can provide effective means for ensuring universal access to the infrastructure and services of the Information Society.'*

The Scenario in Developing Countries

Western transition models from monopoly to competition were 'exported' to poorer countries, though one might wonder whether their validity is universal when national conditions differ so profoundly (Castelli et al., 2000).

In high-teledensity economies (the ITU defines teledensity as the number of main telephone lines per 100 inhabitants), the reform mainly aimed at introducing dynamism in the sector and harnessing the deployment and usage of new services. Far from it, the enforcement of a telecommunications policy in least developed countries is more complex. When networks do not exist, their creation is obviously the first and necessary prerequisite, as any project is necessarily based on the development of an infrastructure. Melody (1997: 20) pointed out that *'it is perhaps a misnomer to consider telecom reform in developing countries as a process solely of reform [...] Their task surely involves reform, but the major objective is to build a national telecom system from the beginnings of the system that the PTOs now provide.'*

In some cases, especially in 'second world' countries, the liberalisation and privatisation of the telecommunications sector has contributed to a more dynamic development.[2] The number of lines has grown substantially during the past few years. In particular, the number of wireless connections, as in other parts of the world, has been subject to impressive growth rates. Nevertheless, even in these cases, the task of achieving an authentically universal service is still far from being achieved.

However, a majority of low-income countries are facing two difficult challenges: attracting foreign capital to subsidize their network construction and, especially, finding ways to compensate the reduction in one of their main financing channels brought on by the modification of the rules regulating international communications. Traditionally, as a matter of fact, their telecommunications industry profits have been insufficient and heavily dependent from the 'net settlement payments'[3] they receive to terminate (that is to convey on domestic networks) incoming international telephone services. The relevance of the international dimension of funding universal service for developing countries is made explicit by countries whose ratio of net settlement payments to total telecommunications revenue in a year can be greater than 20% or 30% and can reach the 50% (Castelli et al., 2000).

The pressure exerted by developed countries (and especially the United States), added to the development of technological alternatives avoiding the usage of traditional operator networks, has led to a reform of this system, basically translated into a dramatic reduction of the payments for each transnational communication (see Thuswaldner, 1998; Stanley, 2000). Those economies are thus much more worried about how the traditional system is breaking down, their investment programmes and the possibility of their operators' viability being jeopardized by the pressures exerted to reduce prices on international services.

Incentives for the Support of a Global Telecommunication Infrastructure Development

As we have said previously, telecommunications access has been a major target of governments during the last century in practically every country in the world. This suggests that the benefits of a large connection to telecommunications services have been perceived regardless of the political option in power. Could the benefits of a broad connection not limited to national frontiers be perceived as well?

Consideration of Externalities

Any telecommunications service presents two types of positive externalities: 'external ones', that appear outside the service itself favourably influencing other productive activity sectors,[4] and 'internal ones' (linked to their own consumption) that result from being network-based activities.

The general definition on why *club* externalities exist is quite simple:[5] since telecommunications networks provide interaction between all users, each new subscriber benefits from (and is prepared to pay for) accessing a group of pre-existent users, whilst offering a new possibility for communication (actual or potential) to that group of connected customers. These 'social' benefits are not taken into account by the individual user when considering the possibility of joining the network. It can occur that the additional benefits the existing customers would receive should the 'marginal' customer join the network (maybe discouraged by a costly subscription fee, not necessarily above costs), exceed the losses the company would incur – if it should reduce the subscription fee to attract that customer. However, it is not easy to include in the network the benefits provided by a new user: there may be many potential beneficiaries, but not all of them can know each other and, even if that were so, it would be difficult to reach agreements. Additionally, those transaction costs could exceed the benefits provided by the externalities (Littlechild, 1979).

Some authors also consider network externalities those that result from the fact that users who do not initiate communications also benefit from a certain utility despite not having paid for the service (Bar & Munk Riis, 1997; Cave et al., 1994).

In new services, the club characteristic is extraordinarily strengthened. With the telephone, the group one interacts with is basically limited to personal or work-related circles, with a highly improbable chance of communicating with 'strangers'. However, whoever enters nowadays in a chat room, an interactive game or a forum does not know most of the time any details of their interlocutor, maybe not even their nationality. The group of users receiving some sort of actual usefulness by the connection of a new member is, thus, impossible to define in advance although, surely, it is much more important than with traditional services.

A second type of network externalities are those considered 'indirect'. Individual usefulness is not only a direct consequence of the number of users, but an indirect one as well, since it also depends on the amount of services available, which represent a portfolio that grows in parallel to the number of users that allow to achieve a return on them.[6]

Telecommunications 'as a Tool'

In the previous paragraphs we have referred exclusively to communicating. However, each individual or institution connecting to a network can also, in addition to communicating, make public all sorts of information, which takes us to the next argument: advanced services are a 'necessary tool' for the enjoyment of other goods.

The basic idea is the one considering telecommunications as a tool for the dissemination of global public goods. An international public good is a benefit-providing utility that is – in principle – available to everybody throughout the globe (Morrissey et al., 2002).

The first of these public goods would be information, or from a broader perspective, knowledge. Knowledge is a global public good because the marginal cost of a new individual receiving it, is zero, while its advantages are geographically unlimited; although some sort of exclusion, which would transform it into an impure public good, is possible, it would not be desirable due to that absence of marginal cost (Stiglitz, 1999). But, and here is the role of the tool, for a country, the adaptation and creation of new knowledge is as essential as its dissemination, which is affected by the effectiveness of its communications system.

The importance telecommunications services have at present, and will further have in the future, for information access, exchange, generation and dissemination, seems without any doubt undeniable. Using Conceição's methodology (2003), we could establish that the underuse of this knowledge would be caused by access problems specified in the underprovision of adequate connection resources.

Second, we must consider the relationship between telecommunications and economic development. Poverty has the property of a public 'bad'. If poverty were to reach even more excessive proportions, it could result in a rising number of failing states, civil strife, international conflict, and international terrorism and crime (Kaul & Le Goulven, 2003). Thus, to reduce extreme poverty can be considered relevant to the goals of global public goods. Also, it is unanimously accepted that any future economic development shall not be viable without advanced telecommunications.

Item 9 of the WSIS *Declaration of Principles* makes this instrumental, although key, role of the ICTs for generating economic growth very clear: *'we are aware that ICTs should be regarded as tools and not as end in themselves. Under favourable conditions, these technologies can be a powerful instrument, increasing productivity, generating economic growth, job creation and employability and improving the quality of life of all.'*

Conclusions

Financing advanced telecommunications services infrastructure requires more than just money. The state's role lies also in providing incentives to enable private actors to contribute to network deployment. Governments should take action in order to support an enabling and competitive environment for the necessary investment in ICT infrastructure and for the development of new services (Item 9 of the WSIS *Plan of Action*). Yet in many countries international financial assistance is absolutely necessary.

The declaration on international cooperation included in the WSIS *Declaration of Principles* is extremely vague: *'we recognize the will expressed by some to create an international voluntary Digital Solidarity Fund, and by others to undertake studies concerning existing mechanisms and the efficiency and feasibility of such a Fund'* (Item 61). The *Plan of Action* dedicates a major section to the Digital Solidarity Agenda although it takes no steps forward in respect of creating mechanisms, and simply promises a review of the adequacy of all existing financial mechanisms, including the feasibility and the creation of the voluntary Digital Solidarity Fund.

Thus, the future of said voluntary fund[7] depends on the generosity of the richer states. However, there is more than a risk that resource allocations will fall short of required funds. It has been argued that the pattern of aid-giving is dictated by political and strategic considerations (Alesina & Dollar, 2000) or even that nation-states are likely to consider spending on international cooperation only if it is in their national interest (Kaul & Le Goulven, 2003).

Therefore, we believe that the only path to success starts with the conviction of possible donors that they are making investments instead of providing a philanthropic contribution. There is no doubt as to the fact that equity provides solid arguments for international cooperation, possibly the most solid ones. However, from a strictly pragmatic point of view it seems necessary to find other reasons. This is precisely what the results of many other programmes traditionally guided by equity-related considerations advise: global inequity is increasing and poverty is still pervasive. Keeping the Digital Solidarity Fund under the 'aid' umbrella would probably lead to equally poor results. Approaching it as a 'cooperation' action would be more adequate. The rationale for aid is equity, while that of cooperation is efficiency. An improvement of efficiency would generate non-restricted benefits, perceived by all the participants of the Information Society.

Arguments used to back plans for the development of enhanced telecommunications infrastructures are almost always too vague. Frequently, their positioning is based on the resource to using scarcely rigorous terms such as 'social importance', 'digital divide' or 'budgetary realism'. The awareness of the role of advanced telecommunications services as a necessary tool for the provision of global public goods and the existence of important externalities would consolidate the convenience and need for those programmes. Specifically, there is the convenience of and need for a Digital Solidarity Fund, which in any other case would probably be relegated to the limbo of appealing but hollow ideas.

Notes

1 *'Any Member has the right to define the kind of universal service obligation it wishes to maintain; such obligations will not be regarded as anti-competitive per se, provided they are administered in a transparent, non-discriminatory and competitively neutral manner and are not more burdensome than necessary for the kind of universal service defined by the Member'.*

2 China, Vietnam, Botswana, El Salvador, Jamaica, Hungary, Mauritius, Chile, The Philippines and Morocco are the 10 countries that moved up the most positions during the 1990–2000 decade in the classification of countries per total telephony density (ITU, 2002).

3 The system ruling settlement procedures in international telecommunications emerged when national monopoly carriers provided international services and has remained fairly static for more than a century. To provide switched telephone services between country A and B, an international carrier of country A must agree with an international carrier of country B upon the terms and conditions. Such compensation, averaged on a 'per minute' basis, is referred to as the 'accounting rate'. Assuming that the international transmission link is jointly owned, a country A carrier owes to a country B carrier one-half of the agreed bilateral accounting rate to terminate a minute of service in carrier B's country. This latter charge is referred to as the 'settlement rate'.

4 Telecommunications services provide an alternative to physical transportation, reduce the transaction costs and contribute to promoting competitiveness. See Gómez Barroso and Pérez Martínez (2003).

5 The pioneer works in telecommunications are those of Artle and Averous (1973) and Rohlfs (1974).

6 See Curien (1993). Jebsi (1997) declares that there is a virtuous circle connecting services and users: more users will lead to the creation of more services, which will attract more users, and so on. Katz and Shapiro (1985) provide a general review of this type of externalities and add post-sales service, information securing and even psychological benefits ('bandwagon effect').

7 The voluntary nature of the fund rules out any options (whose acceptance would indeed be extremely difficult) imposing procedures considered coercive or involving other actors, such as that of Hayashi (2003), proposing a 'global universal service fund' fed by the carriers of richer regions.

References

Alesina, A., Dollar, D. 2000. 'Who gives aid to whom and why?', *Journal of Economic Growth* 5(1): 33–63.

Artle, R., Averous, C. 1973. 'The telephone system as a public good: static and dynamic aspects', *Bell Journal of Economics and Management Science* 4: 89–100.

Bar, F., Munk Riis, A. 1997. 'From welfare to innovation: toward a new rationale for universal service', *Communications & Stratégies* 26: 185–206.

Bardzki, B., Taylor, J. 1998. *Universalizing universal service obligation: a European perspective.* 26th Telecommunications Policy Research Conference. Alexandria, 3–5 October.

Castelli, F., Gómez Barroso, J. L., Leporelli, C. 2000. 'Global universal service and international settlement reform', *Vierteljahrshefte zur Wirtschaftsforschung* 69(4): 679–694.

Cave, M., Milne, C., Scanlan, M. 1994. *Meeting universal service obligations in a competitive telecommunications sector.* Report to European Commission DG IV. Luxembourg: Office for Official Publications of the EC.

Conceição, P. 2003. 'Assessing the provision status of global public goods', pp. 152–179 in Kaul, I. Conceição, P., Le Goulven, K., Mendoza, R.U. (eds.) *Providing global public goods.* New York: Oxford University Press.

Curien, N. 1993. 'Économie des services en réseau: principes et méthodes', *Communications & Stratégies* 10: 13–30.

Falch, M., Anyimadu, A. 2003. 'Tele-centres as a way of achieving universal access – the case of Ghana', *Telecommunications Policy* 27(1–2): 21–39.

Gómez Barroso, J. L., Pérez Martínez, J. 2003. 'Análisis del fundamento económico de una posible ampliación del servicio universal de telecomunicaciones', pp. 145–154 in Joyanes Aguilar, L., González Rodríguez, M. (eds.) *Congreso Internacional de Sociedad de la Información y el Conocimiento. Libro de actas.* Madrid: McGraw-Hill.

Hayashi, T. 2003. *Fostering globally accessible and affordable ICTs.* Report of the ITU 'Visions of the Information Society' project. Downloaded from www.itu.int/osg/spu/visions/papers/accesspaper.pdf.

International Telecommunication Union (ITU). 2002. *World telecommunication development report 2002.* Reinventing telecoms. Geneva: International Telecommunication Union.

ITU. 1998. *World telecommunication development report 1998.* Universal access. Geneva: International Telecommunication Union.

ITU. 1994. *The changing role of government in an era of telecom deregulation.* Report of the

Second Regulatory Colloquium held at the ITU Headquarters 1–3 December 1993. Geneva. Downloaded from http://www.itu.int/itudoc/osg/colloq/chai_rep/2ndcol/coloq2e.html.

Jebsi, K. 1997. 'Effet club, externalité de services et tarification de l'accès au réseau', *Communications & Stratégies* 25: 45–59.

Katz, M. L., Shapiro, C. 1985. 'Network externalities, competition, and compatibility', *American Economic Review* 75(3): 424–440.

Kaul, I., Le Goulven, K. 2003. 'Financing global public goods: a new frontier of public finance', pp. 329–370 in Kaul, I. Conceição, P. Le Goulven, K., Mendoza, R. U. (eds.) *Providing global public goods*. New York: Oxford University Press.

Littlechild, S. C. 1979. *Elements of telecommunications economics*. London: The Institution of Electrical Engineers.

Melody, W. H. 1997. 'Policy objectives and models of regulation', pp. 13–27 in Melody, W. H., editor, *Telecom reform. principles, policies and regulatory practices*. Lyngby: Den Private Ingeniørfond, Technical University of Denmark.

Morrissey, O., te Velde, D. W., Hewitt, A. 2002. 'Defining international public goods: conceptual issues', pp. 31–46 in Ferroni, M., Mody, A. (eds.) *International public goods: incentives, measurement, and financing*. Boston: Kluwer Academic Publishers.

Rohlfs, J. 1974. 'A theory of interdependent demand for a communications service', *Bell Journal of Economics and Management Science* 5: 16–37.

Stanley, K. B. 2000. 'Toward international settlement reform: FCC benchmarks versus ITU rates', *Telecommunications Policy* 24(10–11): 843–863.

Stiglitz, J. E. 1999. 'Knowledge as a global public good', pp. 308–325 in Kaul, I. Grunberg, I., Stern, M. A. (eds.) *Global public goods*. New York: Oxford University Press.

Thuswaldner, A. 1998. 'International telephony revenue settlement reform', *Telecommunications Policy* 22(8): 681–696.

World Summit on the Information Society (WSIS). 2003a. *Declaration of Principles*. Document WSIS-03/GENEVA/DOC/4-E. Downloaded from http://www.itu.int/dms_pub/itu-s/md/03/wsis/doc/S03-WSIS-DOC-0004!!PDF-E.pdf.

World Summit on the Information Society (WSIS) 2003b. *Plan of Action*. Document WSIS-03/GENEVA/DOC/5-E. Downloaded from http://www.itu.int/dms_pub/itu-s/md/03/wsis/doc/S03-WSIS-DOC-0005!!PDF-E.pdf.

10: PSB as an Instrument of Implementing WSIS Aims

BARBARA THOMASS

Introduction

The World Summit on Information Society (WSIS) in Geneva in December 2003 was a key event in questions of global communication. The issue of information thus became a focal point in the global public sphere. But traditional media as print, radio and television did not play an important role in the discussions and final papers, despite WSIS' far-reaching and ambitious aims concerning the role of information, its dissemination and its role for development.

The member states of the UN decided in Geneva to enforce informational rights of men and women (i.e. free access to media and digital services). In 2015 everyone shall have access to radio and television, 50 percent shall have access to the Internet. The media as a whole should work for the enlightenment of society and be enforced as an integrating force for a global vision of free communication. Therefore mechanisms of financing support should be developed.

I want to look in this chapter at the role Public Service Broadcasters can play in its capacity to become an instrument for implementing WSIS aims. Therefore I will look at:

- The notion of information and knowledge;

- Decisions and declarations in the documents concerning traditional media literally;

- Fields of interest in the WSIS documents affecting traditional media;

- The provisions of PSBs for implementing WSIS aims;

- Examples from PSBs of the Western world in doing so;

- Obstacles preventing PSBs from doing so further;

- Perspectives for WSIS 2005 in Tunis.

The main hypothesis of this contribution is: Public Service Broadcasters with their obligation to serve cultural and social purposes are qualified to make an essential contribution to those ambitious aims which are connected to the concept of information and knowledge society.

The Notion of Information and Knowledge

If we consider the content of the World Summit more profoundly, we can state that one of the key issues within the development of the information and knowledge society deals with the suppliers, i.e. the actors who provide knowledge. Common knowledge is still distributed via the traditional audio-visual media. I here refer to the notion of common knowledge used by Gripsrud (1999) who considers television as *'the central medium for the production and mediation of knowledge'*, as *'primary contributor to common knowledge'*. He states that it is the problem of this medium, that elites mostly ignore this function of television. Common knowledge can as well be considered as popularized knowledge; that is as *'widely shared pool of information and perspectives from which people shape their conceptions of self, world and citizenship'* (Gripsrud, 1999: 2). Thus a world declaration, which claims:

> to build a people-centred, inclusive and development-oriented Information Society, where everyone can create, access, utilize and share information and knowledge, enabling individuals, communities and peoples to achieve their full potential in promoting their sustainable development and improving their quality of life. (WSIS Declaration of Principles A)

cannot ignore traditional media. And an action plan, based on this declaration should take print, radio and television into consideration. How did they do so?

Decisions and Declarations in the Documents Concerning Traditional Media Literally

The principles reaffirm the importance of any medium for the reception and impartation of information and ideas and call to recognize the role of the media:

> *We reaffirm*, as an essential foundation of the Information Society, and as outlined in Article 19 of the Universal Declaration of Human Rights, that everyone has the right to freedom of opinion and expression; that this right includes freedom to hold opinions without interference and to seek, receive and impart information and ideas through any media and regardless of frontiers. Communication is a fundamental social process, a basic human need and the foundation of all social organization. It is central to the Information Society.

> *We are resolute* in our quest to ensure that everyone can benefit from the opportunities that ICTs can offer. We agree that to meet these challenges, all stakeholders should work together to: improve access to information and communication infrastructure and technologies as well as to information and knowledge; build capacity; increase confidence and security in the use of ICTs; create an enabling environment at all levels; develop and widen ICT applications; foster and respect cultural diversity; recognize the role of the media; address the ethical dimensions of the Information Society; and encourage international and regional cooperation. We agree that these are the key principles for building an inclusive Information Society. (WSIS Declaration of Principles A)

In a special section, freedom of information, diversity of media and the important role of traditional media in all their forms for the Information Society are underlined:

9) Media

55. We reaffirm our commitment to the principles of freedom of the press and freedom of information, as well as those of the independence, pluralism and diversity of media, which are essential to the Information Society. Freedom to seek, receive, impart and use information for the creation, accumulation and dissemination of knowledge are important to the Information Society. We call for the responsible use and treatment of information by the media in accordance with the highest ethical and professional standards. Traditional media in all their forms have an important role in the Information Society and ICTs should play a supportive role in this regard. Diversity of media ownership should be encouraged, in conformity with national law, and taking into account relevant international conventions. We reaffirm the necessity of reducing international imbalances affecting the media, particularly as regards infrastructure, technical resources and the development of human skills.

The action plan is clear about the promotion of the joint use of traditional media and new technologies and sees traditional media responsible for supporting local content development. Access to traditional media is seen as one important element for the maintenance of cultures and languages and local communities and for facilitating their communication:

C2. Information and communication infrastructure: an essential foundation for the Information Society (Plan of Action)
l) Encourage and promote joint use of traditional media and new technologies.
C8. Cultural diversity and identity, linguistic diversity and local content
e) Support local content development, translation and adaptation, digital archives, and diverse forms of digital and traditional media by local authorities. These activities can also strengthen local and indigenous communities.
f) Provide content that is relevant to the cultures and languages of individuals in the Information Society, through access to traditional and digital media services. [...]
j) Give support to media based in local communities and support projects combining the use of traditional media and new technologies for their role in facilitating the use of local languages, for documenting and preserving local heritage, including landscape and biological diversity, and as a means to reach rural and isolated and nomadic communities. (my emphasis)

Especially in the paragraph dedicated to the media the action plan states that media should be encouraged to play an essential role in the information society and that traditional media should be encouraged to bridge the knowledge divide:

Media

24. The media – in their various forms and with a diversity of ownership – as an actor, have an essential role in the development of the Information Society and are recognized as an important contributor to freedom of expression and plurality of information.

a) Encourage the media – print and broadcast as well as new media – to continue to play an important role in the Information Society.

b) Encourage the development of domestic legislation that guarantees the independence and plurality of the media.

c) Take appropriate measures – consistent with freedom of expression – to combat illegal and harmful content in media content.

d) Encourage media professionals in developed countries to establish partnerships and networks with the media in developing ones, especially in the field of training.

e) Promote balanced and diverse portrayals of women and men by the media.

f) Reduce international imbalances affecting the media, particularly as regards infrastructure, technical resources and the development of human skills, taking full advantage of ICT tools in this regard.

g) Encourage traditional media to bridge the knowledge divide and to facilitate the flow of cultural content, particularly in rural areas.

Fields of Interest in the WSIS Documents Affecting Traditional Media

But there are many fields of interest within the key issues in the WSIS documents, which can as well be considered to affect traditional media. Those are:

- Infrastructure of information and knowledge;
- Digital divide;
- Access to information and knowledge;
- Capacity-building;
- Cultural identity and diversity;
- International and regional cooperation.

I will come back to this later.

The Provisions of PSB for Implementing WSIS Aims

Why is PSB in this contribution declared to be a good tool for implementing those aims? The sense and idea behind these declarations of the international community clarify that they agreed on a formula according to which media are seen not only as a market good, but that they should serve certain social purposes and be orientated to the common good.

After years of experience with a commercialized mediascape we have learned that the implementation of media objectives that focus on the common good and public service is dependent on organizational characteristics of the media. Europe has generated an institution, which has a long-standing experience in serving the public with media content: Public Service Broadcasting, and exported it to the world.

Public Service Broadcasters with their obligation to serve cultural and social purposes are qualified to carry into effect those ambitious aims that are connected to the concept of the information and knowledge society and that also lie behind the event of the World Summit. Furthermore, they are grounded on established and tried-and-tested organizational patterns able to integrate heterogeneous interests concerning the use of information media via controlling and regulating structures. Thus the Public Service Broadcaster can be seen as one of the actors appropriate to supply information in an information society based on a societal consensus.

Looking at public broadcasting as a form of regulation (as Syvertsen does) implies three essential conditions (Syvertsen, 2003: 156):

- Broadcasters serving the public are protected to a certain – varying degree – against market forces, e.g. by securing their financial base.

- They are obliged to serve some fundamental social or cultural aims and purposes that lay beyond consumer's interests. Those obligations are fixed within their licences or special laws.

- To secure those privileges and obligations, certain controlling mechanisms, based on the participation of different social groups and interests, are involved. They work within a social consensus about the content of the obligations.

Thus PSBs dispose of competences, structures, content and so forth, which should be used for the implementation of the ambitious aims described in the WSIS *Declaration of Principles* and WSIS *Plan of Action*. The following is an overview of those topics of the World Summit where PSB can be used as an instrument. Here I refer to the concept of popularized or common knowledge.

Examples from PSBs for Implementing WSIS Aims

Infrastructure of Information and Knowledge
The dissemination of knowledge via television and radio is an important element if we consider the notion of popularized or common knowledge. Public broadcasters have developed a wide range of formats to present knowledge based on all sciences and to do so for many different tastes and educational levels. In their presentation, they are less vulnerable to market forces and can follow their own agenda and programme mission. In world regions with low alphabetization rates and low Internet access rates, radio (and especially community radio) becomes a central element in the dissemination of knowledge and information.

Digital Divide
Some PSBs work hard on broadening the access to the Internet via the use of digital TV. For example the British government is including the BBC into its strategy to move on to a knowledge-based society. Labour is considering the transition to digital TV as a key issue in the development of an information/knowledge society, in order to provide a majority of the British people with Internet access. Offering free high-quality content on Digital TV is therefore a cornerstone of the strategy to draw audiences to digital TV. The BBC took on this

challenge. With the support and backing of politics, the BBC has pushed forward the development of digital techniques, especially digital services and platforms (e.g. its website which is well accepted worldwide), developing interactive services and data applications. Thus the activities of a public broadcaster here serve to overcome the digital divide within British society, as it draws new audiences to the digital services. Elements on the website of the BBC do help people to get involved and become accustomed to the Internet.

Access to Information and Knowledge

Public broadcasters dispose over a huge amount of audio-visual documents concerning history, society, and sciences... in many areas of interests. Although there are many copyright problems, those documents can be considered as a stock of information and knowledge which should be open to the public, as it has been produced (to a high extent) with public funds, i.e. the licence fee or other public financial resources. Giving access to those archives of audio-visual material is possible. Many broadcasters are working on it. For instance, the CBC is concentrating on putting documentaries about important events of Canadian history on the Internet; and the INA in France is developing a database of the archives of France Télévision, which can be used on the spot.

Capacity-Building

The BBC offers a big portal giving access to the many different forms of education and adult education. Anyone interested in improving his or her capacities finds on the BBC website a variety of programmes, supplied by different providers and institutions aimed at individual and professional capacity-building. Something similar can be found at the Deutsche Welle, which also serves as a market place offering professional education. Many PSB TV programmes with relevant content give access to their material via the Internet.

Cultural Identity and Diversity

PSB is organized, in general, in a way that it is providing content for different ethnicities, cultural identities and diverse tastes and opinions in a given society. It withstands (more or less successful) to the pressure of programme mainstreaming. Some PSBs even try to re-enter the road of public service qualities in this way (e.g. the CBC). And they have – more than commercial media – the means and the obligation to serve the aims of strengthening cultural identity and diversity.

International and Regional Cooperation

EBU is a cooperation network, which has already expanded to countries outside of Europe, for instance some Arab countries – thus enforcing and deepening exchange and cultural diversity. Thus cooperations between PSBs throughout the world could be encouraged and deepened, also with the perspective to serve the information and knowledge society.

Obstacles Preventing PSBs From Implementing WSIS Aims

The main condition for PSBs to go further on this road providing knowledge and information beyond the immediate TV programme is dependent on their ability and possibility to use new online media for their purposes. Expanding to this area means expanding their remit.

In many societies media politics, facing the constant pressure of commercial broadcasters, are not willing to give PSBs this possibility. In Germany, for example, Internet activities of ARD and ZDF are strictly bound to their programmes. Any further offerings are – according to the broadcasting law – not compatible with their mission. In the United Kingdom, where this obstacle does not exist, the BBC became the biggest information provider through its portal. Also in Canada is CBC's website the most important provider for information on the Internet. This is even more important as the Canadian information and TV market is heavily flooded by material originating from another culture, i.e. the U.S.

Perspectives for WSIS 2005 in Tunis

These ideas should illustrate that PSB should be brought into the debate as an important tool for the implementation of the aims of the WSIS. This argument can as well be included into the current evaluation of the WSIS achievements and into the preparation of the follow-up *Plan of Action*. The 2003 *Declaration of Principles* and the *Plan of Action* were not very clear about structures, preconditions and tools for pursuing the ambitious aims. Pointing to the provisions PSBs represent, this gap could be closed a little bit further.

References

Gripsrud, Jostein (ed.) 1999. *Television and Common Knowledge*. London, New York: Routledge.

Syvertsen, Trine. 2003. 'Challenges to Public Television in the Era of Convergence and Commercialization', *Television and New Media* 4(2): 155–175.

World Summit on the Information Society (WSIS). 2003a. *Declaration of Principles*. Building the Information Society: a global challenge in the new Millennium, Document WSIS-03/GENEVA/DOC/4-E, 12 December 2003, downloaded on 1 October 2004, http://www.itu.int/dms_pub/itu-s/md/03/wsis/doc/S03-WSIS-DOC-0004!!MSW-E.doc.

World Summit on the Information Society (WSIS). 2003b. *Plan of Action* Document, WSIS-03/GENEVA/DOC/5-E, 12 December 2003, downloaded on 1 October 2004, http://www.itu.int/dms_pub/itu-s/md/03/wsis/doc/S03-WSIS-DOC-0005!!MSW-E.doc.

Afterword: Towards a Knowledge Society and Sustainable Development: Deconstructing the WSIS in the European Policy Context

PETER JOHNSTON
(Head of Evaluation and Monitoring Unit, Information Society and Media DG)

The year 2004 was one of transition in the EU. Firstly to a wider Union of 25 member states, with a greater diversity of interests and levels of development; secondly, to a new Parliament and Commission, both of which will wish to re-orient policies; and thirdly, to the new planning perspectives for the period from 2007 to 2013. The follow-up to the first phase of the World Summit on the Information Society must fit in with these transitions.

The 'Prodi Commission' has already set out the general orientations for the period beyond 2006 (EC, 2004a). These set three new priorities, the first of which is sustainable development through higher growth and better jobs. This is complemented by the third priority for Europe to become a stronger 'global player', notably as a sustainable development partner for the developing world. These priorities have been reflected in the Commission's proposals for the 2nd phase of the WSIS (EC, 2004b).

There is now wide recognition that information and communication technologies are one of the most important contributors to growth and sustainable development. In some countries, notably Ireland and Finland, ICT investment has made the major contribution to productivity and growth. In others, such as Italy and Spain, the impact of these technologies has still been small.

This disparity is highlighted in the recent OECD report (Pilat et al., 2002) – and the EITO 2004 report – on ICT, growth and competitiveness. The key conclusion from this observation is that investment in ICTs must be accompanied by investment in skills and organisational change. We therefore need a more systemic approach to development of a sustainable information society: greater synergy between RTD, regulation and deployment actions; greater investment in more effective public services, notably for health care and education, as well as for administrations; and more active promotion of 'eco-efficient' technologies and their use.

In Europe, the core activity for information society development remains European RTD. This must again be strengthened, and the Commission has proposed five priorities: to realise coherence in the European Research Area; to stimulate increased investment in RTD (to 3% of GDP by 2010); to increase the

European investment in IST; to strengthen the dissemination and exploitation of results; and to show stronger European leadership in global initiatives.

A good example of such leadership exists in the connection of universities into high-speed collaboration network – the 'GEANT' network now covers over 3000 universities and R&D centres in 36 European countries. It interconnects to the U.S., Canada, China, Japan and Korea, and to South America via Brazil. It is now the world-leading research network on which global knowledge exchanges can be built.

Policy and programme re-orientation needs to be built on evidence-based evaluation of the effectiveness of current interventions. We have therefore carried out, in 2004, a mid-term evaluation of the 'eTEN' support to pan-European information infrastructures; a five-year assessment of IST research and technology development; and an independent study of how all these measures contribute to the 'Lisbon' and 'Sustainable Development' Strategies.

These evaluations will all feed into the reviews and re-orientations of the Lisbon and Sustainable Development Strategies in early 2005, by the new Commission.

The 'Digital Europe' project has been a key element in linking our activities on the information society to sustainable development. There are six major links:

- Higher 'added value' in all products and services;

- Some products become immaterial services;

- More efficient supply chains and transport logistics;

- Improved energy efficiency in intelligent buildings and vehicles;

- More efficient use of buildings and city infrastructures (EC, 2004c);

- A better 'work-life' balance through use of ICT – with more work in local communities and better land-use planning.

We can therefore see the following issues emerging in the WSIS and its follow-up: to strengthen the link between the 'information society transition' and 'sustainable development':

- A clear causality between effective ICT-use and innovation-led growth: ICT as a key factor in development;

- The e-Europe Action Plan as a model for sustainable national e-strategies throughout the world;

- Recognition of the important role of the private sector, not just as suppliers of ICT, but in promoting effective use (through CRS and the Global eSustainability Initiative);

- A new focus on the 'digital divide', both in Europe's regions and worldwide: access to knowledge-infrastructures for learning and entrepreneurship; and

- New initiatives for resource efficiency: eco-efficient technologies, and resource efficiencies through innovative uses of new technologies.

These issues are elaborated in the Commission's proposals for the 2nd phase of the WSIS-COM (2004) 480 of 13.07.04 (EC, 2004b). I commend them to your attention.

References

European Information Technology Observatory (EITO). 2004. Frankfurt: EITO.

European Commission. 2004.101, February 2004.

European Commission. 2004. 480, 13 July 2004.

European Commission. 2004. 60.

Pilat, D., Lee, F., Van Ark, B. 2002. 'Production and use of ICT: A sectorial perspective on productivity growth in the OECD area', *OECD Economic Studies* 35(2): 47–78.

Recommendations on the Subject of Research and Education in the Area of the Information Society

ECCR[1], January 2005

The Information Society, as a concept and a vision, is the driving force of a major shift in communication and information management. ECCR acknowledges the decisive role played by the European institutions and the Commission in particular, together with other international organisations such as OECD, the World Bank, the Unesco and ITU, not only to promote, but also to shape and map out a mainspring of European development.

However, there are clear signs that the IS is loosing momentum and has now reached a decisive crossroads. The initial vision, which drove the first and spectacular phase of ICT development, led to a model based predominantly on technology and commerce, which did not live up to the expectations. Evidence suggests that implementation of ICTs will lead to a mature and desirable Information Society only if certain conditions can be met, and challenges be faced, not in discourse but in facts:

- Bridging the digital divide (1): access to ICTs should be made possible not necessarily to everybody indistinctively, but especially to those who are underprivileged.

- Bridging the digital divide (2): giving access to technologies is worthless unless a matching effort is undertaken in education so as to level up the users' skills and ability to make efficient and responsible use of these technologies, not only to find and retrieve relevant content (including local content), but also to produce and make available their own content.

- Internet governance: although the Internet embodies a certain vision of freedom, the Information Society cannot be left to the law of the strongest, nor can it be regulated by particular interests, be they of a nation or an industry.

- Enhancing democracy: the emerging technologies must determinedly serve the advent of democracy and, in already democratic regimes, feed a process of revival of political institutions and citizen participation beyond mere governmental websites or fancy e-voting.

Europe needs a new, clearer and carefully thought vision, which can be referred to in innovating, implementing, using and regulating the Information or the

Knowledge Society in the making. ECCR believes that this can only be achieved through an increased and redeployed effort in *research* and *education*, in consultation with the academic community. Given the complexity of the issues, efforts to structure and sustain academic *networking* initiatives are to be increased.

Research is excessively concentrated in the areas of technological innovation and market development, both areas feeding each other in a circular relationship, with a prevailing priority on short-term return on investment and industrial applications. Meanwhile there is an endemic deficit of research aiming at solutions to identified problems within a broader societal perspective. As a result, there is an urgent need for a sizeable effort to undertake or revitalise research in neglected areas, promoting social research not in addition, but in close connection with industrial research from the earliest stages of development.

In full accordance with the principle of subsidiarity, and given the intrinsic transnational nature of the Information Society in the making, the European institutions are to enhance efforts and activity in high-level research, with a particular emphasis in the following perspectives:

- Scientific research, along with policy-making, are to develop beyond mere market regulation and development to *encompass the social aspects* of communication in the broadest sense, focusing specifically on the users, their expectations, their fears, their needs; studying the *social and cultural implications* of the Information Society.

- The perspectives on the Information Society are to be broadened beyond the spectacular, yet restrictive questions of innovations associated with the Internet and mobile communication to *include all vectors of information flows* including traditional media and the entertainment industry.

- In particular, there is an urgent need to examine *the role of public service and community radio and television* in Europe and to determine the way in which it can balance the rapid evolution of private broadcasters towards a certain vision of media content driven by the sole concern of attracting audiences.

- European authorities are to *establish clear standards of indicators* to monitor the various aspects of the development and implementation of the Information Society and carry out the measures and analysis thereof.

Sound policy and more generally harmonious development of societies in Europe require extensive, transdisciplinary, transnational and long-term research efforts involving the scientific community and in close connection with civil society, the industry and political institutions, thus amplifying the participatory processes initiated within the World Summit for the Information Society (WSIS), for the benefit of all.

Regarding the funds allocated by the European Union to scientific research, we acknowledge the efforts of the Commission to support the academic research community in a context where other sources of funding, particularly that of national governments, are lacking dramatically. We acknowledge also the latest improvements, which can be found in the 6th Framework Program for Research and Development. We regret, however, the lack of transparency in determining the priority topics covered by the programme, and we call also for a thorough reorganization of the evaluation process which, in its current form, has been a massive source of misunderstanding, of missed opportunities and, ultimately, of a loss of motivation.

Education efforts are to be developed dramatically. Current initiatives are meagre and concentrate on the acquisition of computer skills with an overwhelming focus on tasks-oriented tools and procedures, falling short of providing even the minimal foundation needed to orient oneself in the Information Society in the making. The severe deficit of adequate education leads to a new form of illiteracy, which entails societal risks comparable to that of illiteracy of the past centuries. This deficit is just as dramatic as regards media literacy efforts with children as well as with adults, which remains in no way proportionate to the role that media have taken as a prime source of information, culture and leisure.

Just like the Information Society should be considered in a broader perspective, the education deficit is to be framed within the pre-existing shortage of media education at large. The scarce attention given to media in educational systems is in complete discrepancy with the prevailing role played by television in particular and increasingly by Internet and video games, in shaping people's access to information as well as their sociability at large.

ECCR recommends that the European Commission actively encourages a structured and systematic approach to critical media literacy at all levels in a similar way that it encourages the development of other basic skills such as command of foreign languages.

Note

1   See also http://www.eccr.info.

Notes on Contributors

Michel BAUWENS (michelsub2003@yahoo.com) is the founder of the Foundation for Peer to Peer Alternatives, a virtual organization and network publishing the weekly newsletter Pluralities/Integration. He has been active as an entrepreneur creating two dotcom companies in Belgium, as well as having been the eBusiness Strategy Manager for the leading telecom company Belgacom, and European Thought Manager for US/Web CKS. He has been teaching the Anthropology of Digital Society for the Facultés St-Louis/ICHEC in Brussels, Belgium, co-edited two books with the same title (with Salvino A. Salvaggio) and co-produced with director Frank Theys, a three-hour TV documentary on the metaphysics of technology, entitled TechnoCalyps, the Metaphysics of Technology and the End of Man. He currently teaches Globalization at Payap University in Chiang Mai, Thailand. Apart from the research into the nature of P2P processes, he is also producing a world-centric and participative bibliography on the past, present and future of human civilization.

Bart CAMMAERTS (b.cammaerts@lse.ac.uk) is a political scientist and media researcher working at Media@LSE, London School of Economics & Political Science, UK. In 2002 he obtained a PhD in social sciences at the Free University of Brussels with a thesis bearing the title: *Social Policy and the Information Society: on the changing role of the state, social exclusion and the divide between words and deeds*. After that Bart Cammaerts did post-doctoral research at ASCoR (University of Amsterdam). He researched the impact of the Internet on the transnationalization of civil society actors, on direct action and on interactive civic engagement. Currently he lectures on Citizenship and Media at the LSE and holds a Marie Curie research fellowship. In this capacity he studies the use of the Internet by international organizations (UN and EU) in order to involve civil society actors in their decision-making processes and its effects on the ground in terms of networking and democratising global or regional governance processes.

Nico CARPENTIER (Nico.Carpentier@kubrussel.ac.be) is a media sociologist working at the Communication Studies Departments of the Catholic University of Brussels (KUB) and the Free University of Brussels (VUB). He is co-director of the KUB research centre CSC and member of the VUB research centre CEMESO. His theoretical focus is on discourse theory, his research interests are situated in (media)domains as sexuality, war & conflict, journalism, (political and cultural) participation and democracy. In 2004 he edited the book *The ungraspable audience* and wrote together with Benoît Grevisse *Des Médias qui font bouger*. Since 2004 he has been a member of the Executive Board of the ECCR.

Miyase CHRISTENSEN (miyaseg@yahoo.com) is an Assistant Professor in the Faculty of Communication at Bahcesehir University in Istanbul. Dr Christensen obtained her PhD from the University of Texas at Austin in 2003 with a doctoral dissertation on the relationship between Turkish and European Union telecommunications and IST policies. In addition to her work on Turkey and the

EU, she has written on digital divide, U.S. telecommunications policy and Turkish media and national identity.

Claudio FEIJÓO GONZÁLEZ (cfeijoo@gtic.ssr.upm.es) holds a PhD in Telecommunications Engineering from the Universidad Politécnica de Madrid. Currently he is an Associate Professor at ETSI Telecomunicación de Madrid and Co-ordinator of the Telecommunications Regulation Group (GRETEL). Dr. Feijóo has participated in different public and private research projects for the main organisations and companies of the Spanish Information and Communications Technologies sector. He has also been Special Adviser for the State Secretariat of Telecommunications and Information Society as well as reviewer of R&D projects for the European Commission. His present interests include the development and prospective of new services and infrastructures and their regulation.

Divina FRAU-MEIGS (meigs@wanadoo.fr) teaches American Studies and Media Sociology at the ParisIII–Sorbonne University and specialises in media and information technologies of Anglo-Saxon countries. She is also a researcher at the CNRS (Social Uses of Technology), editor in chief of *Revue Française d'Etudes Américaines* (RFEA) and member of the editing board of *MédiaMorphoses* (INA-PUF). She has written many publications about the media (*Les Ecrans de la Violence*, Economica, 1997; *Jeunes, Médias, Violences*, Economica, 2003), technologies and screen subcultures (*Médias et Technologie: l'exemple des Etats-Unis*, Ellipses, 2001) and the connection between media and technologies (*Le crime organisé à la ville et à l'écran, 1929–1951*, Armand Colin, 2001; *Médiamorphoses américaines*, Economica, 2001). She currently works on the subjects of cultural diversity and acculturation through the study of real format programmes (reality shows) from an intercultural perspective (Big Brother). She is also interested in the research of media regulation and self-regulation. At present she is vice-president of the International Association for Media and Communication Research (IAMCR) and she has been a board member of the French Information and Communication Studies Society (SFSIC) and the European Consortium for Communication Research (ECCR).

José Luis GÓMEZ BARROSO (jlgomez@cee.uned.es) holds a PhD in Economics from the Universidad Nacional de Educación a Distancia (UNED). He also holds a degree in Telecommunication Engineering from the Universidad Politécnica de Madrid as well as another degree in law from the Universidad Complutense. Currently he is an Assistant Professor in the Applied Economics Department at Universidad Nacional de Educación a Distancia (UNED). His research interests are focused on telecommunications economics and regulation as well as on public aspects of the development of Information Society. He is a member of the Telecommunications Regulation Group (GRETEL).

Ana GONZÁLEZ LAGUÍA (agonzalez@gtic.ssr.upm.es) gained her Telecommunications Engineer degree from the Universidad Politécnica de Madrid. Currently, she is a PhD student at the Research Group in

Communications and Information Technologies (GTIC-SSR-UPM). Her interests focus on the influence of ICT convergence on regulatory aspects and the politics of telecommunications, especially spectrum issues. She is also a member of the Telecommunications Regulation Group (GRETEL).

Peter JOHNSTON (peter.Johnston@cec.eu.int) is responsible for evaluation of Information Society policies and programmes in the European Commission. He has worked with the Information Society (and Media) DG of the European Commission since 1988. He has been responsible for the strategic planning of European telecommunications research (the RACE and ACTS programmes), and helped prepare the 5th Framework Programme. He has also had responsibility for EC actions in the area of telework stimulation, electronic commerce, multi-media access to cultural heritage, and for sustainable development in a knowledge economy. Dr Johnston has wide experience in international research co-ordination: from 1976 to 1984, he worked at the OECD, and from 1984 to 1988 he was responsible for research on pollution control in the UK Department of Environment. He read physics at Oxford University, and was a Fulbright-Hays scholar at Carnegie Mellon University, and at Oxford University until 1976.

Stefano MARTELLI (martelli@unipa.it) is Full Professor of Sociology of Culture and Communications at the University of Palermo (Italy). Martelli is a member of the ECCR Board of Advisors, and member of many other international and national sociological associations. He has written a wide range of books and papers, which were presented at national and international conferences. His most recent (co-authored or co-edited) books are *Comunicazione multidimensionale. I siti Internet di istituzioni pubbliche e imprese* (2003, Franco Angeli, Milano), *Il Giubileo "mediato". Audience dei programmi televisivi e religiosità in Italia* (2003, Franco Angeli, Milano) and *Immagini della emergente società in rete* (2004, Franco Angeli, Milano).

Claudia PADOVANI (claupad@libero.it) is researcher of Political Science and International Relations at the Department of Historical and Political Studies at the University of Padova, Italy. She teaches International Communication and Institutions and Governance of Communication, while conducting research in the fields of the global and European governance of the information and knowledge society. She is particularly interested in the role of civil society organizations and transnational social movements as stakeholders in global decision-making processes. From this perspective she has followed closely the WSIS process and has written extensively on the experience. She is a board member of the International Association for Media and Communication Research (IAMCR) and of the international campaign Communication Rights in the Information Society (CRIS).

Sergio RAMOS VILLAVERDE (sramos@gtic.ssr.upm.es) gained his Telecommunications Engineer degree from the Universidad Politécnica de Madrid. He is a PhD student and a member of the Research Group in Communications and Information Technologies (GTIC-SSR-UPM). His interests focus on the technological, economic and regulatory aspects of 3G mobile communications systems. Sergio Ramos has worked for a consulting firm specialising in access providers operators. Currently he is Resident Twinning Adviser on regulatory issues for the Latvian government. He is also a member of the Telecommunications Regulation Group (GRETEL).

David ROJO ALONSO (drojo@gtic.ssr.upm.es) is a PhD student at the Research Group in Communications and Information Technologies (GTIC-SSR) of the Universidad Politécnica de Madrid (UPM). His main interests focus on policy and regulatory aspects of telecommunications sector, particularly the process of transposition of the new European regulatory framework to the Spanish legislation, and its future impact on the telecommunications market. At present, he is also a member of the Telecommunications Regulation Group (GRETEL).

Ned ROSSITER (n.rossiter@ulster.ac.uk) is a Senior Lecturer in Media Studies (Digital Media) at the Centre for Media Research, University of Ulster, Northern Ireland, and Adjunct Research Fellow at the Centre for Cultural Research, University of Western Sydney. Ned Rossiter is co-editor of *Politics of a Digital Present: An Inventory of Australian Net Culture, Criticism and Theory* (Melbourne: Fibreculture Publications, 2001) and *Refashioning Pop Music in Asia: Cosmopolitan Flows, Political Tempos and Aesthetic Industries* (London: RoutledgeCurzon, 2004). He is also a co-facilitator of *fibreculture*, a network of critical Internet research and culture in Australasia www.fibreculture.org.

Jan SERVAES (j.servaes@uq.edu.au) is Professor and Head of the School of Journalism and Communication at the University of Queensland in Brisbane, Australia; Editor-in-Chief of *Communication for Development and Social Change: A Global Journal* (Hampton Press), Associate Editor of *Telematics and Informatics: An international journal on telecommunications and Internet technology* (Pergamon/Elsevier), and Editor of the Hampton Book Series *Communication, Globalization and Cultural Identity*. He has taught International Communication and Development Communication in Belgium (Brussels and Antwerp), the USA (Cornell), The Netherlands (Nijmegen), and Thailand (Thammasat, Bangkok). From 2000 to 2004 he was President of the European Consortium for Communications Research (ECCR) and Vice-President of the International Association of Media and Communication Research (IAMCR), in charge of Academic Publications and Research.

Trained as a teacher, Bart STAES (bstaes@europarl.eu.int) has already been active for more than 20 years in European politics, first as a political group assistant, then as MEP. He is presently MEP for Groen! (the Flemish Greens) and one of the most active members of the Flemish delegation in the European Parliament. Bart Staes is best known for his work for more transparency and anti-fraud measures in the Budgetary Control Committee. He is also a substitute member of the Committee on the Environment, Public Health and Food Safety and member of the delegation to the EU-Russia Parliamentary Cooperation Committee. Bart Staes was one of the shadow draftsmen on the Regulation establishing a European Food Safety

Authority. He is also very concerned with peace, human rights (Chechnya) and the rights of minorities (Kosova).

Barbara THOMASS (Barbara.Thomass@rub.de) is Professor at the Ruhr-University Bochum, Institute for Media Studies, for International Comparisons of Media Systems. Her main research interests and publications have been media politics in Western Europe, media and journalism ethics, media systems in Eastern Europe, regional television in Europe, and the future of public service broadcasting. Before her academic career she worked as a journalist.

Arjuna TUZZI (arjtuzzi@stat.unipd.it) is researcher of Political Science at the Department of Historical and Political Studies (University of Padova, Italy). She has a PhD in Applied Statistics for Economics and Social Sciences and teaches Methods for the Social Research and Statistics for the Social Sciences. Her main research interests concern content analysis and statistical analysis of texts; closed-ended questionnaire, open-ended interview and the problem in comparing different stimuli-answers data collecting tools; statistical methods in evaluation; electoral data analysis and Italian electoral abstensionism; multimedia, long-life learning and online education.

Paul VERSCHUEREN (paul.verschueren@skynet.be) is a PhD candidate from the School of Journalism and Communication at the University of Queensland, Australia. He has an MA in Linguistics and an MA in Social and Cultural Anthropology from the Catholic University of Leuven, Belgium. His research interests include the cultural dimensions of media technologies, visual representation, and ethnography. His doctoral work focuses on the visual framing of conflict and violence.